Palgrave Studies in Environmental Sustainability

The series will take advantage of the growing interest in a number of environmental issues in sustainability, particularly in those that focus on interdisciplinary approaches to solving real-world problems. Unfortunately, many of the themes of sustainability address short- and long-term survivability of our planet and there is a pressing need for access to information to those who seek to solve sustainability problems. Thus, the scope of the book series seeks to be comprehensive within an environmental sustainability framework. The topics covered in the series range from climate change to public land management. While no series can ever be truly comprehensive, the series will provide one of the most definitive surveys of knowledge in the area of environmental sustainability.

Robert E. Mace

Groundwater Sustainability

Conception, Development, and Application

palgrave macmillan

Robert E. Mace
The Meadows Center for Water and the Environment
Texas State University
San Marcos, TX, USA

Palgrave Studies in Environmental Sustainability
ISBN 978-3-031-13515-6 ISBN 978-3-031-13516-3 (eBook)
https://doi.org/10.1007/978-3-031-13516-3

Cover Credit: Tatsiana Volskaya

This Palgrave Macmillan imprint is published by the registered company Springer Nature Switzerland AG.
The registered company address is: Gewerbestrasse 11, 6330 Cham, Switzerland

"The water that comes from beneath the surface of the earth has from time immemorial been regarded by man as a mysterious boon—cold, clear, and presumably pure; wonderfully refreshing to the hot and weary wayfarer; essential, perhaps, to the very life of the community, yet without cost and generally regarded as inexhaustible; coming from some dark orifice in the earth, its source unknown and unknowable; a great blessing freely and mysteriously bestowed on man by some benevolent but capricious Providence. When, very rarely, in times of unusual drought, this mysterious blessing, which was regarded to be perennial and inexhaustible as the air itself, was cut off and the spring or well went dry, the occurrence was a calamity, in the face of which man stood helpless in fear and awe."

Oscar E. Meinzer (1920)

for Wendy
and all the springs I've loved before

Acknowledgments

I am deeply thankful to Geary Schindel with the Edwards Aquifer Authority for recommending me as a possible author for this book and Robert Brinkman, series editor, and Palgrave Macmillan for the opportunity to write it. I thank Vanessa Puig-Williams and Maurice Hall of the Environmental Defense Fund; Larry French, Cynthia Ridgeway, and John Dupnik of the Texas Water Development Board; Van Kelley of INTERA; David Yoskowitz of Texas A&M University-Corpus Christi; Quinn McColly of Texas Water Trade; and many others for discussions on groundwater sustainability over the past year (and before). I am grateful for the review comments of two anonymous reviewers and those of Louis (Trey) W. Mixon, III of BlueTriton Brands Inc./Ozarka, Ronald Green of Southwest Research Institute, and Regina Allen of the Meadows Center on Water and the Environment on a draft of this book. I am also grateful to Quinn McColly of Texas Water Trade for a review of the economics chapter, Julie Lewey of the Devils River Conservancy (and now the Nueces River Authority) for a review of the section on Val Verde County, and John Dupnik of the Texas Water Development Board (and former general manager of the Barton Springs Edwards Aquifer

Conservation District) for a review of the section on the Barton Springs Segment of the Edwards Aquifer. I also thank the Meadows Foundation and the Cynthia and George Mitchell Foundation for supporting my general work on groundwater sustainability and climate change, some of which appears in the book. All errors and omissions are solely mine.

Contents

List of Figures

1

The Importance of Groundwater Sustainability

"Are we being good ancestors?"
Jonas Salk (*1992*)

When Thomas Patten Day of Fort Worth, Texas, sunk a borehole 100-meters deep into the rock beneath his land in 1876 (Hill 1892, p. 96; Marsh and Marsh 1993), he probably had no idea that his well would inspire the drilling of hundreds more across the Blackland Prairies of Texas as well as the rest of the state (Fig. 1.1). Thousands of wells had already been sunk in the Lone Star State, but these wells were hand dug or shallow, needing pails and ropes or windmills and sucker rods to raise the water. Patten's well was different: water rose in the borehole of its own volition and freely flowed onto the land surface. Not only was this magical but also, at a time when pump technology was limited, it was invaluable: a well that raised its own water! Furthermore, protected from human activities by thick layers of clay and shale, the water quality was impeccable and consistent, unlike the local rivers and streams polluted with human waste and animal carcasses.

The people of Fort Worth took note, and soon there were a hundred flowing wells in town sunk to about the same depth (Hill 1892). Almost

R. E. Mace, *Groundwater Sustainability*, Palgrave Studies in Environmental Sustainability, https://doi.org/10.1007/978-3-031-13516-3_1

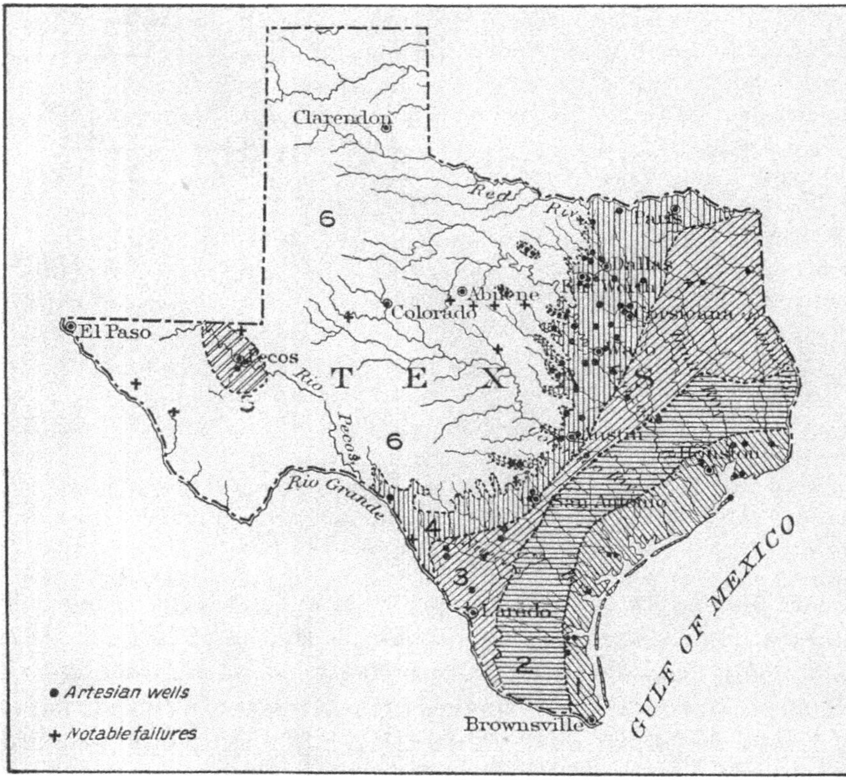

Fig. 1.1 Location of artesian groundwater zones in Texas circa 1900 (from Hill 1901, p. 395). The area with the vertical lines is the Blackland Prairies. Hill (1901) did not include all flowing artesian wells on this map

guaranteed of success, farmers and ranchers west of town also began to sink hundreds of wells into the artesian aquifer below.

Robert T. Hill, the father of hydrogeology in Texas, noted that multiple sandy outcrops about 10 to 50 kilometers west of Fort Worth were the source of the water and suggested drilling into deeper sandstones with potentially greater flows beneath Fort Worth and the Blackland Prairies from an aquifer named Trinity. As a result of this recommendation, in 1884, Fort Worth sunk a 1000-meter-deep well just south of downtown on Tucker's Hill, a well that pushed water 20 meters above the street and flowed an astounding 32 liters per second (Fort Worth Gazette 1891).

Bequeathed with the name Jumbo, the deep, Fort Worth well caught the attention of more people across the state. The state government drilled several wells on the capitol grounds in Austin and communities across the prairies, people across the state, and brewers (upon the discovery of artesian water, Adolph Busch, among others, established breweries fed by artesian water) drilled deep holes resulting in hundreds more wells sunk into deeper sands. Drillers tapped sands and limestones with successful wells in Dallas, Denton, Taylor, Austin, McGregor, Belton, Waco, San Antonio (Fig. 1.2), and many other communities.

Because of their success in drilling flowing wells, Fort Worth became known as the City of Artesian Wells, Waco as Geyser City, and the Blackland Prairies as the Great Artesian Basin. Robert T. Hill, in a report to the United State Congress, called it the "Artesian Epoch" of Texas.

Successful flowing wells across the region resulted in unbridled exuberance about the supply, what it meant for the economic prospects of Texas, and unusual theories about the source of the water (Mace 2016 in the next five paragraphs unless otherwise noted). A judge, E.G. Bower, declared the supply as sourcing from the Ozark Mountains in southern Missouri and northern Arkansas. Some thought the water came from the Great Lakes (Dallas Morning News 1890) while others believed that the source of the water was snowmelt from the Rocky Mountains. An imaginative Professor C.B. Wilbur of the Nebraska Academy of Sciences declared that all of Texas was underlain by a 30-meter-thick sand connected to the limitless supply of the Arctic Ocean. Robert T. Hill correctly identified the source as the sandstone outcrops just west of Fort Worth and Waco, but most locals and local luminaries ignored his scientifically derived conclusions.

Regardless of where the water came from, all parties, including Robert T. Hill, thought the aquifers were limitless in their supply and that "the ordinary use of it as it would be to empty a lake with a ladle" (Dallas Morning News 1893). Hill believed that the sandstones captured 50 percent of all the rainfall that fell on them (later studies would reveal it was about 1 percent). When asked if all the new wells would decrease the supply, Hill replied "Theoretically, yes; practically, never." and went on to conclude that "Dallas has in these wells an inexhaustible supply of water."

Fig. 1.2 Artesian wells in San Antonio, Texas circa 1895 (Hill and Vaughan 1898, p. 278)

Fort Worth projected that its wells could supply a bustling city of a million people. Dallas, after sinking its first artesian well off the southwest corner of its courthouse in 1890—The Dallas Geyser—reveled in its flow of 30 to 45 liters per second, prompting the mayor to call for 100 more wells drilled within two years with the goal of having a well on every corner of town (Dallas Morning News 1891). An industrious local proposed the drilling of 270 wells to direct the flow into the nearby Trinity River to turn Dallas—about 350-kilometers inland—into a "seaport city."

Perhaps prompted by promises of an inexhaustible supply and the status of having one, people who owned flowing artesian wells let them flow unabated whether they were using the water produced by them or not, creating rivers on the streets as they flowed downhill into creeks and rivers. Rural users often directed artesian flows into a wellhouse where they kept perishables under the cool water for refrigeration and preservation. By 1892, nearly 2200 liters per second of water was gushing out of the artesian aquifers underlying the Black and Grand prairies between Del Rio, San Antonio, Austin, Waco, Dallas, and Forth Worth (Dallas Morning News 1892).

The future for flowing artesian water supplies looked great in the Great Artesian Basin of Texas, but then hydrogeologic reality set in. As early as 1891, well owners began to observe a decrease in flow where wells were drilled close to each other. Government scientists measuring water pressures in the wells noticed widespread pressure declines over time, indicating non-sustainable use of the resource. Their frantic calls for shutting off wells when not in use were ignored, people preferring to believe the "science" that accommodated their actions rather than the science which required a change in behavior.

The flows in hilltop wells were the first to decline to a trickle and then to nothing at all, yet people at lower elevations continued to let their wells flow uncontrolled. By 1894—less than 20 years after Thomas Patten Day's discovery well—about 237 of 240 wells drilled in Fort Worth had stopped flowing. In 1890, a prominent real estate company in Dallas stated that "the continual flow of artesian wells is too uncertain to be relied on for a permanent supply." By 1903, many of the artesian wells drilled across the Blackland Prairies had ceased to flow. Despite calls for the state to protect the flowing wells of the Great Artesian Basin, the legislature did not pass laws preventing the waste of artesian flows until 1913, almost two decades after flowing artesian wells started to fail. The great Artesian Epoch of Texas was over almost as soon as it began.

Although this is a story specific to Texas, it is a story about groundwater that has been repeated many times across the world. The plot line is almost always the same: discovery, irrational exuberance, and overexploitation followed by belated (and often ineffective) regulation. Yet each story is unique due to the properties of the aquifers, the foundational laws and legal opinions, and the willingness of users and society to manage their groundwater resources. Some aquifers regulate themselves either through low productivity, which limits use to small users, or through being highly responsive to use, thus affecting its use within years rather than decades. Most aquifers, however, take decades or longer to respond to production, enticing use to far exceed sustainable levels and resulting in environmental, economic, and social impacts. In these cases, achieving sustainable groundwater use is the most challenging, since reducing groundwater use to sustainable levels may affect the livelihoods of people (generally farmers) and have broader economic and societal impacts.

The Importance of Groundwater

Groundwater is a critical source of water to the world, and for good reason—it is often accessible, abundant, and affordable. Of all the water on Earth, only about 2.5 percent—35 million cubic kilometers—is freshwater, and only 31.3 percent of that—10.67 million cubic kilometers—is not ice or snow in the Arctic and Antarctic (these and subsequent numbers are based on Shiklomanov 1993). Of that remaining freshwater, 98.7 percent of it—10.53 million cubic kilometers—is groundwater. Lakes are about 0.85 percent of all surface water on the planet while rivers are a mere 0.02 percent (Fig. 1.3).

While volume of water in storage is important, so is how much water is moving into and out of these water categories. Of all the precipitation (such as rain, snow, and sleet) that falls on the continents in an average year—119,000 cubic kilometers—about 61 percent (72,000 cubic kilometers) is evaporated back into the atmosphere, 38 percent (44,800 cubic kilometers) runs off the land surface and into rivers, and 1 percent (2200 cubic kilometers) replenishes aquifers (Fig. 1.4). Annual average flows into the planet's rivers are about 20 times greater than flows into its

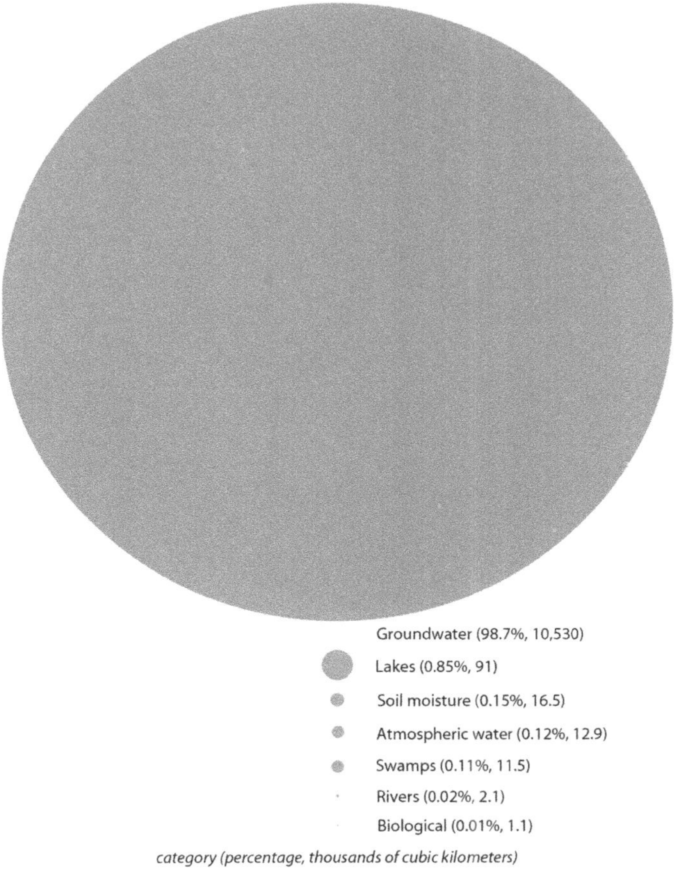

Fig. 1.3 Volumes of fresh, unfrozen water on Earth (data from Shiklomanov 1993; note that percentages do not add up to 100 percent due to rounding)

aquifers. The volume of freshwater in storage in the planet's aquifers is about 4800 times greater than the amount of flow into those aquifers and 235 times greater than the flows in the planet's rivers. Rivers do not store much water, but they have a lot of water moving through them. Groundwater systems store a lot of water, but not a lot of water moves through them compared to surface water flows and compared to groundwater storage.

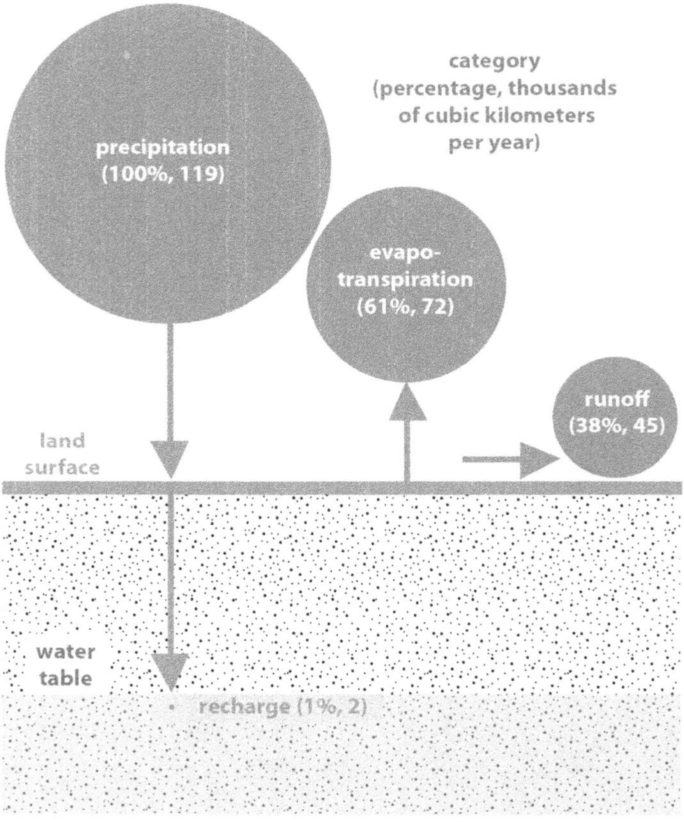

Fig. 1.4 Volumes per year of precipitation and how much precipitation evapotranspirates, runs off, and recharges aquifers (data from Shiklomanov 1993)

A simple water-budget approach suggests that the highest the sustainable development of groundwater could be is the annual amount that groundwater is replenished,[1] but there is a huge amount of water stored in the aquifers. This is what I call the Paradox of Groundwater Availability: a massive amount of water in storage with a small amount available for annual sustainable use (<< 0.02 percent of the total volume in storage). This paradox is what makes the sustainable management of groundwater

[1] As we will learn later, sustainable development is generally much lower than this.

politically difficult except in certain circumstances, such as aquifers that respond quickly to development or where governments seek sustainable management before users exceed sustainable levels of production. Those that seek to maximize the economic benefits of groundwater today point to the massive volume of water in storage while those concerned about water-level declines, springflows, and sustainability point to the dramatically smaller amount of natural recharge.

Despite the relatively low amounts of water flowing in aquifers, groundwater is an important part of the water cycle, albeit, in many cases, a long, slow detour where water may travel hundreds to thousands of years underground before re-emerging at the Earth's surface. That re-emergence—sometimes dramatic through large springs but oftentimes unnoticed through countless seeps—provides baseflow to streams and rivers, supporting ecosystems and other water users. In the United States, groundwater is thought to contribute as much as 40 percent of the average annual amount of water flowing in the country's streams and rivers (Alley and others 1999).[2] Groundwater also commonly contributes water to lakes, wetlands, estuaries, and continental margins. Shallow groundwater also supports a special category of plants called phreatophytes where deep roots tap into shallow aquifers.

Aside from natural lakes, societies build reservoirs to store surface water—they do not have to build reservoirs for groundwater because aquifers are themselves reservoirs. However, to access groundwater, outside of capturing springflows, people must sink wells and then produce water from those wells. Humans first sunk wells by hand, perhaps after watching animals dig in stream bottoms for water or perhaps by following seasonal springs and seeps into the subsurface. These wells became more sophisticated over time and were lined with stone or brick to access deeper parts of the subsurface. Eventually, humans developed deep drilling techniques through the patient chipping of rock to reach thousands of meters into the ground followed by the mud-rotary techniques of today.

[2] Note that this number is not in harmony with the Shiklomanov (1993) numbers, although, admittedly, the Shiklomanov (1993) numbers are for the planet while the Alley and others (1999) analysis is for the United States. Nonetheless, if the Shiklomanov (1993) number for runoff held true in the United States, then groundwater flows would be about 29,900 cubic kilometers per year instead of 2200.

To produce water from these wells—unless they were free-flowing—humans first used plant or animal materials, such as the sewn-together leaves of plants or harvested kidney bladders. People eventually developed buckets and Archimedes screwpumps powered by animal or wind. Eventually, modern windmills accessed deeper water until fossil-fueled and electric-powered centrifugal pumps allowed large amounts of production from shallow and deep aquifers.

Because of availability and human ingenuity, groundwater now supports 26 percent of human water use, about 982 cubic kilometers per year as estimated for 2010 with 70 percent for irrigation, 21 percent for domestic uses, and 9 percent for industrial use (numbers in this paragraph are from Margat and van der Gun (2013) unless otherwise noted). Groundwater provides 27 percent of water for North America, 14 percent for Europe (including the Russian Federation), 21 percent for Africa, 30 percent for Asia, 14 percent in South America, and 25 percent for Oceania. Much of that groundwater supports crops through irrigation. In 2010, 70 percent of the groundwater used across the globe went to agriculture with 13 percent in Africa, 76 percent in Asia, 43 percent in Oceana, 36 percent in Europe (including the Russian Federation), 50 percent in South America, and 70 percent in North America. About 40 percent of all cropland irrigation is sourced from groundwater (Siebert and others 2010). Groundwater is also an important supply of potable water, providing 209 cubic kilometers per year (Margat and van der Gun 2013) to more than two billion people on the planet (Morris and others 2003; Foley and others 2011). With climate change expected to decrease surface-water resources, groundwater may become an alternative supply (Foster and Tyson 2019), thus placing additional stresses on the world's aquifers.

Unfortunately, much of the groundwater produced to support human activities has been produced unsustainably. Estimated groundwater depletion from 1900 to 2008 is about 4500 cubic kilometers, averaging about 145 cubic kilometers per year from 2000 through 2008 (Konikow 2011). This groundwater depletion volume is about 0.04 percent of the fresh groundwater in storage, and about 6.6 percent of annual flows into groundwater systems (comparing to numbers from Shiklomanov 1993). While these percentages might not seem large, water-level declines in the world's groundwater systems have caused storage declines in many of the largest aquifers on most continents (Famiglietti 2014), decreased spring-flows by as much as 6.6 percent (based on the analysis above), decreased groundwater flows into wetland regions by as much as 75 percent (Döll and others 2019), caused as much as 77 percent of the world's land subsidence (Bagheri-Gavkosh and others 2021), caused sea-water intrusion (Bear and others 1999), and impacted the sustainability of the world's agricultural output (Aeschbach-Hertig and Gleeson 2012). In addition to the local and regional effects described above, groundwater depletion has resulted in raising sea levels by 12.6 millimeters (Konikow 2011) and changing the Earth's spin axis (Adhikari and Ivins 2016).

Increased groundwater use inevitably led to conflict between groundwater users, although it is unclear exactly when this began. Ancient groundwater law, which reaches back to at least Greek times (Drummond and others 2004), held that groundwater was owned along with the land, not recognizing that groundwater moves in the subsurface and that it is connected to neighboring pieces of land and to springs and rivers (Utton 1982). This ancient treatment of groundwater tied to property carried forward into Spanish Law and English Common Law which still echoes today in the laws of many nations (Utton 1982).

As people drilled more and more wells, especially after World War II when centrifugal pumps and affordable power allowed large amounts of groundwater to be produced from willing aquifers, more and more wells interfered with each other or with springs and rivers, resulting in lawsuits and law-making that, in some cases, modified ownership or required more holistic management that considered all users, interactions between groundwater and surface water, and, yes, managing groundwater sustainably. However, in many places across the planet, especially in developing countries, there is little to no groundwater management (Famiglietti 2014). But where aquifers have no management and little production, there is an opportunity for policy to get ahead of groundwater production and, at the very least, purposefully consider how to best manage limited groundwater resources. Accordingly, sustainable development has become more of a focus as planners and regulators seek to address concerns over the impacts of unsustainable groundwater production.

Purpose

The purpose of this book is to provide a self-contained overview of groundwater sustainability, including its technical terminology, its history, its application, and its pros and cons. While it's difficult—if not impossible—to be all things to all people, I believe that this book will have value to anyone interested in groundwater sustainability. I have attempted to make the book understandable to stakeholders and policymakers by getting only as technical as I need to. Similarly, general experts in sustainability will gain an understanding of the scientific complexities of aquifers and the policy complexities of why groundwater sustainability can be challenging. Although hydrogeologists will find the technical chapter too basic, they will find the historical overview and policy discussions interesting and useful. And finally, for professors and students, I have included questions at the end of each chapter to further discussion of each chapter's concepts and conclusions to challenge students to relate what they have learned to their local area.

When I set out to write this book, I decided to read the literature forward in time from the introduction of safe yield in 1915 to today. I did this to better understand the history and development of the concept of safe yield, how safe yield fits into sustainability, and to better understand the angst that engineers, scientists, and policymakers often have with safe yield and sustainable yield. While reading the literature, I was surprised to find that there is a great deal of misunderstanding by engineers, scientists, and policymakers on the fundamentals of safe and sustainable yield (concepts rooted—by definition—in science and policy and not just science) as well as safe yield itself (a term that has appropriately evolved with time as society has evolved with time). In this sense, I hope this book clarifies the fundamentals and unravels the misunderstandings (although I am sure folks will still disagree on definitions and differences).

Although groundwater sustainability can include water quality (i.e., the sustainability of protecting groundwater quality from human activities that contaminate groundwater), in this book, I focus on the sustainability of groundwater production and groundwater discharge to the surface. I mention water quality as part of this discussion only when induced changes in water quality affect the sustainability of groundwater production, such as saltwater intrusion. Readers interested in the broader topic of groundwater quality and sustainability can find it addressed by Kresic (2008), Maloszewski and others (2012), Zuber and others (2012) as well as others.

Whereas developed countries can address human-caused and natural contamination through water treatment, developing countries may not be able to afford this option. Even in developed countries, rural groundwater users are often responsible for their water quality from their wells (e.g., this is the case in the United States). Furthermore, emerging contaminants such as pharmaceuticals, personal care products, and "forever chemicals" such as per- and polyfluoroalkyl substances (forever because they break down slowly in the environment and in people) are appearing in more and more of the world's water resources, including groundwater.

Ultimately, considering groundwater sustainability requires a multidisciplinary approach of groundwater scientists, sustainability experts, social scientists, economists, policy analysts, facilitators, lawyers, and an educated public. I hope this book weaves this diverse expertise together for enlightened discussions on the future of our groundwater resources.

I Come from Texas!

You may be wondering why someone from Texas should be trusted with writing a book about groundwater sustainability. After all, Texas hosts the international poster child for unsustainable groundwater use, the draining of the Ogallala Aquifer on the Southern High Plains. Furthermore, the foundational legal doctrine in Texas is the Rule of Capture, which allows landowners to pump as much as they want regardless of the impacts to neighbors, springs, and streams. However, the Rule of Capture does not rule Texas, at least from a groundwater governance and management perspective.

In 1949, the Texas Legislature—after a decade-and-a-half-long debate on state versus local control—allowed for the creation of locally controlled groundwater conservation districts. Inside these districts, the management perspective of the Rule of Capture is modified, often substantially, through spacing and permitting requirements. Since 1949, 100 groundwater conservation districts have been created as well as three special districts, the Edwards Aquifer Authority and two subsidence districts in the Houston area.

These 100+ locally controlled groundwater regulating entities sprinkled across a rich variety of hydrogeologic settings and management challenges have created a marketplace of ideas on how to best manage limited groundwater resources. Some districts have engineered their rules to allow, in practice, unfettered use of groundwater while others are managing for sustainable use or are striving for sustainable use. At present, 60 percent of Texas is inside a groundwater regulating entity where 90 percent of all groundwater is produced. And much can be learned about managing groundwater sustainably by studying aquifers that are not managed sustainably.

State technocrats have been calling for the sustainable management of groundwater since the 1960s (although they fell silent after 2002; Mace 2021). In 1968, state water planners even proposed moving water from the Mississippi River across the state to help bring overpumped aquifers into sustainable management.

When I started my job at the state water-planning agency in 1999, my boss sat me down and told me there were two words I could never say in public: (1) metering and (2) sustainability. But a lot has changed over the past 20 years. A number of wells are now metered—including in the Ogallala—and people can talk about sustainability without (generally) being verbally tarred and feathered and then run out of town. As Texas urbanizes and grows, rural areas are becoming more concerned about the development of their groundwater resources as thirsty cities seek new water supplies. California's Sustainable Groundwater Management Act has also heightened interest in Texas for sustainable groundwater use.

Despite some successes, Texas is also a place struggling with sustainability—most of our aquifers are not being produced or planned to be produced sustainably (Mace 2021). But we do have sustainability success stories, albeit driven by federal laws, fears of groundwater export, a simple lack of water or large users, or the aquifer itself. Given my background in Texas and familiarity with local policy issues, I will favor examples from the Lone Star State (write about what you know!), but I have also strived to bring in other national and international examples of groundwater management and sustainability.

Policy and Science

Having worked as a government hydrogeologist for nearly two decades, I quickly learned to recognize the bright line between science and policy. Retaining my job depended on staying on the science side of that line. Academically, it is also important to note where science ends and policy begins. However, many scientists, engineers, and experts are blind to this line, bringing their personal policy opinions to groundwater management discussions along with their professional conclusions on the science. In my opinion, this is a disservice to science as well as to the public and people having to make decisions on groundwater.

Science is the systematic study of the structure and behavior of the physical and natural world through observation and experiment, while policy is a course or principle of action adopted or proposed by a government, party, business, or individual (OUP 2021). Policy, in other words, is a human construct. Science is agnostic on how to manage groundwater, including whether or not it is managed sustainably. *Scientists* may not be agnostic (although if their views affect their science or the perceptions of their science, they are—in my opinion—no longer scientists: they are advocates), but the science is. Science can quantify, with appropriate qualifications, what the outcomes of management schemes might be or, when given management or policy goals, how much groundwater can be produced to achieve those goals. Science is a critical part of establishing and assessing sustainability goals, but science supports policymaking—it does not dictate it.

Given my background and viewpoint, I have written this book to inform, not advocate. I have too much respect for people and policymakers to muddy the groundwaters with personal policy preferences. I also believe muddying the groundwater with personal policy preferences compromises the perception of the science if not compromises the science itself. Scientists, engineers, and experts need to facilitate stakeholders and policymakers in making their own decisions.

I am also human, so I have personal policy preferences and biases. Just so you know where I come from, in case my personal biases unintentionally seep into the book, I believe that we should manage aquifers sustainably where we can. The *where we can* bit comes from the difficulty of putting the pumping cat back in the bag once an aquifer is produced unsustainably. For example, it is difficult for me to imagine that it is politically possible to reduce pumping from the Ogallala Aquifer in Texas 85 percent to achieve sustainability. In addition, if there's a choice between starving a population and managing groundwater unsustainably, perhaps the best choice is to manage unsustainably for a time so people can eat. But for those aquifers where the pumping cat is still in the bag, stakeholders and policymakers should consider sustainability as an option for managing their groundwater.

As I write this in my office at Texas State University, I can turn my head 45 degrees and see San Marcos Springs flowing about 4300 liters per second, a literal river emerging from the ground. At one time, these springs were vulnerable to ever-increasing pumping and drought but now benefit from sustainable management of the aquifer intended to preserve springflow for the endangered species that live therein. San Marcos Springs are not only an oasis in a drouthy part of Texas but also a critical source of water for cultural, recreational, environmental, and downstream uses. All of these uses are preserved through the sustainable management of the aquifer.

Organization

I organized this book to first establish a scientific foundation, then present the developmental history and resulting definitions of safe and sustainable yield, followed by a discussion of groundwater governance and sustainability and several examples. In Chaps. 2 and 3, I discuss how groundwater works to arm you with the terminology and basics of hydrogeology to better understand the issues, challenges, and opportunities for groundwater sustainability. It is easy to oversimplify how much groundwater can be produced to achieve sustainability, but groundwater sustainability is far more complicated.

There are many books out there on hydrogeology. Although I cut my teeth on Freeze and Cherry (1979) and Domenico and Schwartz (1990) when I attended college, I recommend the clear presentation of Fetter (2001) for more ambitious readers, although his book still serves a healthy plate of graphs and differential equations. Since hydrogeologists don't need me lecturing them on hydrogeology, I instead attempted to present a somewhat comprehensive nontechnical description of the basic concepts of groundwater flow. Hopefully I succeeded.

Chapters 4, 5, and 6 focus on defining sustainability and how hydrogeologic experts have addressed groundwater sustainability over the years. I originally planned for one chapter on these topics, but I found the literature and history richer and more nuanced than I expected. Chapter 4 focuses on the introduction of the term safe yield and subsequent discussions of the term. Chapter 5 discusses the introduction of economic analysis and the consideration of groundwater mining instead of safe yield. And Chap. 6 presents the rise of the global discussion of sustainable development that led to greater use of the term sustainable yield. Safe yield and sustainable yield have been controversial topics in the literature and in practice, and people still argue today about what they mean.

Chapter 7 focuses on groundwater governance and management for groundwater sustainability. This includes discussion of legal frameworks and how people, policy, and science fit into governance. Certain legal doctrines are more amenable to sustainable management than others with some doctrines not supporting sustainable development at all. Groundwater governance concerns how decisions on how to manage groundwater are made and how the preferred governance structure does or does not support sustainable development. Groundwater management is how boots-on-the-ground legal doctrines and outcomes from groundwater governance are employed and enforced.

Chapter 8 focuses on pathways to sustainability, including the different types of groundwater sustainability and examples of each type. This chapter will identify reasons for how and why certain areas are managed sustainably.

Finally, in Chap. 9, I present my thoughts on the path to sustainability as informed by my personal experience and the case studies and what happens when we don't manage aquifers sustainably.

Questions

How important is groundwater to your area?
What percentage of water use in your area is from groundwater?
How are water levels responding to groundwater pumping in your area?
Has springflow declined or dried up?

References

Adhikari, S., and Ivins, E.R., 2016, Climate-driven polar motion – 2003–2015: Science Advances, 2:e1501693, 10 p.

Aeschbach-Hertig, W., and Gleeson, T., 2012, Regional strategies for the accelerating global problem of groundwater depletion. Nature Geoscience, v 5, n 12, p 853–861.

Alley, W.M., Reilly, T.E., and. Franke, O.E., 1999, Sustainability of groundwater resources: U.S. Geological Survey Circular 1186, Denver, Colorado, 79 p. https://pubs.usgs.gov/circ/circ1186/.

Bagheri-Gavkosh, M., Hosseini, S.M., Ataie-Ashtiani, B., Sohani, Y., Ebrahimian, H., Morovat, F., and Ashrafi, S., 2021, Land subsidence – A global challenge: Science of the Total Environment, https://doi.org/10.1016/j.scitotenv.2021.146193, 17 p.

Bear, J., Cheng, A.H.-D., Sorek, S., Ouazar, D., and Herrera, I., (editors), 1999, Seawater intrusion in coastal aquifers – Concepts, Methods, and Practices: Kluwer Academic Publishers, 627 p.

Dallas Morning News, October 15, 1890, The Dallas Geyser.

Dallas Morning News, April 26, 1891, Dallas boring wells.

Dallas Morning News, January 4, 1892, Artesian investigation – Condensation of Prof. R.T. Hill's forthcoming report.

Dallas Morning News, August 24, 1893, Artesianwater – The great lake of it on which Texas rests.

Döll, P., Trautmann, T., Göllner, M, and Schmied, H.M., 2019, A global-scale analysis of water storage dynamics of inland wetlands – Quantifying the impacts of human water use and man-made reservoirs as well as the unavoidable and avoidable impacts of climate change: Ecohydrology, 18 p., https://doi.org/10.1002/eco.2175.

Domenico, P.A., and Schwartz, F.W., 1990, *Physical and Chemical Hydrogeology*: John Wiley & Sons, 824 p.

Drummond, D.O., Sherman, L.R., and McCarthy, E.R., Jr., 2004, The Rule of Capture in Texas – Still so misunderstood after all these years: Texas Tech Law Review, v 37, n 1, 97 p.

Famiglietti, J.S., 2014, The global groundwater crisis: Nature Climate Change, v 4, p 945–948.

Fetter, C.W., 2001, Applied Hydrogeology (4th Edition): Prentice-Hall, 598 p.

Foley, J. A., Ramankutty, N., Brauman, K.A., Cassidy, E.S., Gerber, J.S., Johnston, M., Mueller, N.D., O'Connell, C., Ray, D.K., West, P.C., Balzer, C., Bennett, E.M., Carpenter, S.R., Hill, J., Monfreda, C., Polasky, S., Rockstrom, J., Sheeha, J., Siebert, S., Tilman, D., and Zaks, D.P.M, 2011, Solutions for a cultivated planet: Nature, v 478, 337–342.

Fort Worth Gazette, 1891, Artesian Water-Fort Worth Lays Claim to Being the Geyser City-Three Powerful Artesian Veins Underneath the City-Five Hundred Gallons Per Minute Sixty Feet Above the City: v 15, n 222, edition 1, Monday, May 25, 1891.

Foster, S., and Tyson, G., 2019, Climate-change adaptation & groundwater: Strategic Overview Series, International Association of Hydrogeologists, 6 p.

Freeze, R.A., and Cherry, J.A., 1979, *Groundwater*: Prentice-Hall, Inc., 604 p.

Hill, R.T., 1892, On the occurrence of artesian and other underground water in Texas, New Mexico, and Indian Territory together with the geology and geography of those regions: Final Reports of the Artesian and Underflow Investigations of the Department of Agriculture, Washington, D.C., 166 p.

Hill, R.T., 1901, Geography and geology of the Black and Grand Prairies, Texas, with detailed descriptions of the Cretaceous formations and special reference to artesian waters: Twenty-First Annual Report of the United States Geological Survey to Secretary of the Interior, 1899–1900, Part VII-Texas, Government Printing Office, Washington, 666 p. + plates.

Hill, R.T., and Vaughan, T.W., 1898, Geology of the Edwards Plateau and Rio Grande Plain adjacent to Austin and San Antonio, Texas, with reference to the occurrence of underground waters: *in* Walcott, C.D., (director), Eighteenth Annual Report of the United State Geological Survey to the Secretary of the Interior, 1896–97, Part II—Papers Chiefly of a Theoretic Nature, Government Printing Office, Washington, D.C., p. 193–322.

Konikow, F. 2011, Contribution of global groundwater depletion since 1900 to sea-level rise: Geophysical Research Letters, v 38, 5 p.
Kresic, N., 2008, Groundwater Resources – Sustainability, Management, and Restoration: The McGraw-Hill Companies, Inc., 852 p.
Mace, R.E., 2016, So secret, occult, and concealed – An overview of groundwater management in Texas: in Law of the Rio Grande, Water Law Institute, CLE International, April 14–15, 2016, La Fonda, Santa Fe, New Mexico, Section H3, 27 p.
Mace, R.E., 2021, Five gallons in a ten-gallon hat – Groundwater sustainability in Texas: The Meadows Center for Water and the Environment, Texas State University, Report 2021–08, 50 p.
Maloszewski, P., Witczak, S., and Malina, G., editors, 2012, Groundwater Quality Sustainability: CRC Press, 370 p.
Margat, J., and van der Gun, J., 2013, *Groundwater Around the World – A Geographic Synopsis*: CRC Press (Taylor & Francis Group), 341 p.
Marsh, B., and Marsh, A., 1993, He brought first bathtub to Fort Worth: The Tarrant County Historical Journal, Spring Edition.
Morris, B.L., Lawrence, A.R.L., Chilton, P.J.C., Adams, B., Calow, R.C., and Klinck, B.A., 2003, Groundwater and Its Susceptibility to Degradation – A Global Assessment of the Problem and Options for Management: United Nations Environment Programme, Early Warning and Assessment Report Series, 03-3, 126 p.
OUP (Oxford University Press), 2021, Oxford English Dictionary: https://www.oed.com/.
Salk, J., 1992, Are We Being Good Ancestors? World Affairs: The Journal of International Issues, v. 1, n. 2, p. 16–18.
Shiklomanov, I., 1993, World fresh water resources *in* Gleick, P.H., editor, *Water in Crisis – A Guide to the World's Fresh Water Resources*: Oxford University Press, p 13–24.
Siebert, S., Burke, J., Faures, J.M., Frenken, K., Hoogeveen, J., Döll, P., and Portmann, F.T., 2010, Groundwater use for irrigation – A global inventory: Hydrology and Earth System Sciences, v 14, p 1863–1880.
Utton, A.E., 1982, The development of international groundwater law: Natural Resources Journal, v 22, n 1, p 95–118.
Zuber, A., Maloszewski, P., Witczak, S., and Malina, G., 2012, Groundwater quality sustainability: CRC Press, 364 p.

2

Water Through a Stone

Letting the days go by, let the water hold me down. Letting the days go by, water flowing underground. Into the blue again, after the money's gone. Once in a lifetime, water flowing underground
David Byrne, Brian Eno, Chris Frantz, Jerry Harrison, and Tina Weymouth (*1980*)

Groundwater is mysterious. It lurks in darkness, creeping slowly through the Earth, often for hundreds to thousands of years, before magically appearing as springs and seeps at the land surface. Water is supposed to flow downhill, yet groundwater somehow emerges from the depths through springs and flowing wells, seemingly against gravity. Sometimes the water emerges hot, as if it passed through Hades on its way to the surface, even stinking of sulfur. It's no wonder that the Ohio Supreme Court, in words later quoted by the Supreme Court in Texas, described groundwater as "secret, occult, and concealed."

Early philosophers and scientific thinkers wondered about the source of groundwater (the next five paragraphs are based, unless otherwise noted, on Mace and others [2004]). In the Iliad, Homer (800 to 701 BC) wrote of "the deep-flowing Oceanus, from which flow all rivers and every sea and all springs and deep wells" where groundwater sourced from the

R. E. Mace, *Groundwater Sustainability*, Palgrave Studies in Environmental Sustainability, https://doi.org/10.1007/978-3-031-13516-3_2

oceans, traveled uphill into the interior of continents, and then emerged as freshwater springs that flowed back to the oceans. This Oceanus Theory remained active for about 2500 years with Thales (circa 624 to circa 545 BC), Pliny the Elder (circa 23 to 79 AD), Leonardo da Vinci (1452 to 1519 AD), and Johann Kepler (1571 to 1630) adhering to it.

In his book *Meteorologica*, Aristotle (384 to 322 BC) introduced the Condensation Theory when he stated that "the air surrounding the earth is turned into water by the cold of the heavens and falls as rain … [and]…the air which penetrates and passes the crust of the earth also becomes transformed into water owing to the cold which it encounters there. The water coming from the earth unites with rainwater to produce rivers. The rainfall alone is quite insufficient to supply the rivers of the world with water."

Seneca (4 BC–65 AD) thought that groundwater could come from three possible sources: (a) the Earth itself containing a lot of moisture that is continually being forced out, (b) air within the Earth is continually being converted into water by the forces of darkness and cold (the Condensation Theory), or (c) the Earth is simply being converted to water. René Descartes (1596–1650 AD) revived the condensation theory in the 1600s but with a twist. He thought that ocean water moved into the earth by underground channels where it was vaporized by the heat of the Earth's interior. This vaporized water then rose through caverns, condensed at a higher level, and flowed out of springs.

Vitruvius (circa 80 to 20 BC) presented the beginnings of a correct theory of groundwater when he noted that rain and snow fell on the mountains, percolated through sediments at the foot of the mountains, and later emerged as springs and streams. Bernard Palissy (1509 to 1590 AD), in addition to being an accomplished potter, published a book in 1580 that revived the Vitruvius theory with additional detail: "rain water that falls in the winter goes up in summer, to come again in winter… And when the winds push these vapors the waters fall on all parts of the land, and when it pleases God that these clouds (which are nothing more than a mass of water) should dissolve, these vapors are turned into rain that falls on the ground.… And these waters, falling on these mountains through the ground and cracks, always descend and do not stop until they find some region blocked by stones or rock very close set and condensed. And they rest on such a bottom and having found some channel or other opening, they flow out as fountains or brooks or rivers according to the size of the opening and receptacles."

In the 1600s, a large part of the debate on the source of groundwater was the common belief that there wasn't enough rainfall to account for the flows in springs and streams; therefore, groundwater had to be sourced from something other than rain. In a nifty bit of hypothesis testing, Pierre Perrault (1608 to 1680 AD) measured rainfall in a basin and observed that the volume of rain was about six times greater than streamflow. Later, Edmund Halley (1656 to 1742 AD) measured evaporation rates and calculated that evaporation from the Earth's oceans was enough to account for all the world's water discharged by streams and springs. Edmé Mariotté (1620 to 1684 AD) discussed the percolation theory and showed that increasing and decreasing flows at many springs depended on rainfall.

In 1856, Henry Darcy (1803 to 1858 AD) published his book *Les Fontaines Publiques de la Ville de Dijon* (Public Foundations of the City of Dijon) with an equation, now known as Darcy's Law, that could be used to calculate the volume of groundwater flow at equilibrium if you knew certain aquifer properties. Seven years later, A.J.E. Dupuit (1804 to 1866 AD) used Darcy's Law to derive an equation that described the flow of groundwater to a well at equilibrium. In 1935, C.V. Theis (1900 to 1987 AD) developed the non-equilibrium equation and solution for flow to a well.

Hydrogeologists discovered ways to measure aquifer properties and discovered how aquifers released water. With the advent of computers, groundwater scientists developed numerical groundwater flow and transport models to test our understanding of aquifers and project how water levels and springs might be affected by production and drought.

When considering groundwater sustainability, it's important to know how groundwater works: What are aquifers? What are the different types of aquifers, and how do they differ? How does water get into an aquifer? How does water flow through an aquifer? Where does that water go? How does an aquifer respond to production? In this chapter, I provide the basics of hydrogeology—the science of groundwater—relevant to a discussion of groundwater sustainability.

What Is an Aquifer?

Interestingly, there isn't a single, accepted definition of the word ***aquifer*** beyond the fusion of two Latin words, aqua-/aqui- for water and -fer for bearing resulting in a word that, literally at least, means water bearing. The French physicist François Arago (1786 to 1853 AD) introduced the word aquifer in 1834 in his essay on artesian wells and springs and used it as an adjective meaning water-bearing (Arago 1834). In 1897, William Harmon Norton (1856 to 1944 AD) of the Iowa Geological Survey was the first to use aquifer in English, and he used it as a noun (Norton 1897; Meinzer 1923).

While an aquifer does, indeed, need to be water bearing, there is more to being an aquifer than bearing water. Norton (1897) defined aquifer as a permeable, water-bearing layer. We'll discuss permeability in more detail later, but permeable means "having pores or openings that permit liquids or gases to pass through" (Merriam-Webster 2021). As a result, Norton introduced the ability for an aquifer to not only hold water but also to allow that water to pass through it. However, even the tightest clay layer can transmit water and is more commonly referred to as an aquitard: from the Latin *aqua-/aqui-* (water) and *retardare* (slow).

Later definitions added a nod towards permeability, such as Lohman's (1972) "a formation, group of formations, or part of a formation that contains sufficient saturated permeable material to yield significant quantities of water to wells and springs." Lohman used part of a definition by Meinzer (1923; "a formation, group of formations, or part of a formation that is water-bearing"). While Meinzer removed "permeable" from his definition, Lohman added it back but used "yield significant quantities" to qualify that the formation was permeable enough to (ahem) yield significant quantities. But how much is significant? Other definitions use "usable" or "economically usable" as a qualifier for an aquifer.

Scientifically, "significant quantities," "usable," "economically usable," and other such terms and phrases are unsatisfying because what is significant, usable, or economically usable depends on who—or what—is using it. A sandstone that may not yield enough water for a small town may be enough for a household. And what is significant flow to a spring? Is a seep that provides water to a few plants and a passing dragonfly enough to be considered an aquifer? In the end, what is an aquifer is in the eye of the beholder. One person's aquitard is another person's aquifer. An aquifer can also refer to several aquifers closely associated with each other, although ***aquifer system*** is probably a better term in these cases.

I mention all this because (1) it was kind of fun and (2) to make the point that later, when we discuss aquifers and aquitards, we always need to keep in mind that groundwater is found nearly everywhere and can meet a wide range of needs, including those of the environment. In this book, when we talk about aquifers and sustainability, we will lean toward those aquifers that produce enough water for irrigation and cities, primarily because these are the ones with the greatest amount of use and the greatest sustainability challenges. In some ways, not yielding "significant amounts" of water is a blessing for an aquifer since it may, as a consequence, not be pumped unsustainably, but even aquifers with lower yields can be pumped unsustainably.

Since everyone seems to have their own definition (and even the U.S. Geological Survey doesn't have a single, official definition), I might as well have my own: an aquifer is geologic media that provides water to wells and springs.

Flow Through an Aquifer

One right-of-passage in grade school is learning about the water cycle. When I learned the water cycle, the teacher had us perform a play with students acting as the ocean, evaporation, clouds, condensation, precipitation, runoff, and rivers flowing back to the ocean (I was evaporation). Back in my day, we didn't talk about infiltration, percolation, recharge, groundwater flow, and springs, probably because of that whole secret and occult thing (although I'm happy to report that most students in my university class say that groundwater was part of their grade school experience). Although the volume of water that flows through aquifers is small compared to rainfall, hidden compared to runoff, and—in most cases—slow compared to rivers, groundwater is nonetheless important.

In this section, we will follow a water molecule as it moves into, through, and out of an aquifer and use this as an opportunity to present the different concepts and terms of groundwater flow. Like any science, hydrogeology is full of technical jargon and lingo. In the discussion below, I will *italicize* technical terms needed to follow our water molecule on its journey, but I will ***bold italicize*** critical terms that I will use later in the book. I have also included a glossary at the end of the book that you can refer to while reading later chapters. Entire books, courses, and degrees are dedicated to studying the flow of water in the subsurface. A motivated reader can refer to *Applied Hydrogeology* by C.W Fetter (2001), a particularly well-written introductory book for collegiate beginners.

Before there is groundwater, there has to be a source of water, and that natural source is almost always precipitation, generally in the form of rainfall or snow melt. As we discussed in Chap. 1, only a fraction of what falls from the sky makes it to an aquifer. Our water drop has to first run the gauntlet of interception by plants and *evaporation* from the land surface (not to mention the lovely term, *virga*, which is rainfall that evaporates before hitting the ground). *Infiltration* refers to the movement of water from the land surface into the ground beneath. When the rainfall rate exceeds the infiltration rate, we get *run-off*, the movement of water over the land surface. *Percolation* refers to the movement of water through the subsurface after it infiltrates.

Although some plants can access water through their leaves, most sip water through roots in the soil, carry that water to their leaves, and then release that water as vapor, along with other waste products such as oxygen, through plant pores called stoma. All this photosynthetic exhaling is called *transpiration*. Evaporation and transpiration together are referred to as ***evapotranspiration***, which is the term generally used for these processes, although some will treat evaporation from surface-water bodies such as lakes and rivers separately.

For our water molecule to continue on its journey through the groundwater part of the water cycle, it has to first escape the root zone. The depth of the root zone depends on the soil, the availability of water, and the plants. The Shepherd's Tree in the Kalahari Desert of Africa has the deepest roots ever found at 68 meters (Canadell and others 1996). More typically, roots reach a depth of a few meters (e.g., Foxx and others 1984; Schenk & Jackson, 2002).

As our water molecule moves through the subsurface, it is moving through *pores*—the space between the solid particles that make up soil and sediment. Although pores are not, by definition, required to be connected, they generally are. *Porous media* are geologic materials such as sand, clay, sandstone, limestone, and shale that have pores. The fraction that pores make-up of a volume porous media is called *porosity* (e.g., the porosity of sandstone might be 0.15, or 15 percent of a volume of sandstone). As it infiltrates and then percolates, our water molecule is also sharing pore space with air, which may be connected over long distances between pores or exist as trapped bubbles. Where there is air in the subsurface, it is referred to as the ***unsaturated zone***, which is also called the *vadose zone*.

The thickness of the unsaturated zone and how quickly water travels across it varies depending on the geology. Some aquifers have, in places, no unsaturated zone and others have an unsaturated zone of more than 300 meters. How long it takes a water molecule to traverse an aquifer's unsaturated zone also varies considerably. In limestone aquifers with tall vertical caves, water moves across the unsaturated zone in mere seconds while other aquifers may require hundreds to thousands of years.

The movement of water in the unsaturated zone is controlled in large part by gravity pulling water downward but also *capillarity* and permeability. *Capillarity* refers to the ability for fluids to move in pores due to the forces of adhesion between water and a surface, oftentimes in opposition to gravity (think about slightly dipping a paper towel into water and watching water wick upwards into the paper towel). We will discuss permeability after we reach the aquifer.

Recharge

Eventually, our water molecule will percolate to the ***water table***, which is the top of the aquifer in this location. Water that crosses the surface of the water table is defined as ***recharge***. Although the word ***recharge*** is sometimes used as a general term for any water entering an aquifer regardless of the source, the technical definition—which even many groundwater scientists get wrong—is specific to water crossing the water table. Because recharge is such an important concept when discussing sustainability, my preference is to use the proper technical definition since it is precise (and proper).

Water may cross into an aquifer from other water-bearing formations below or, if the aquifer is faulted, to the sides, or even from the top in certain circumstances. This is referred to as ***cross-formational flow*** since the water is crossing from one formation into another. This is not recharge, although this water may be an important part of the water budget of an aquifer.

Beneath the water table, the geologic media is fully saturated with water, referred to as the *saturated zone*, as opposed to the *unsaturated zone* where water shares the pore space with air. This is also where water in the ground becomes groundwater. ***Groundwater*** is generally restricted to the saturated zone. The literalist in me slightly rebels at this definition since it neglects the water in the unsaturated zone which is also in the ground, but I was not around to formally protest the original definition. I believe that groundwater became groundwater because the initial focus on groundwater was on water in aquifers. As the science of hydrogeology developed and scientists understood that there was also water in the unsaturated zone, it was too late to change the term.

Water Levels

Imagine (or go get!) a glass filled with ice but only partially filled with water (Fig. 2.1). You can think of the ice as porous media, the space between the ice cubes as pores, the level of water in the glass as the water table, the space below your water table as the aquifer or saturated zone, and the space above your water table as the unsaturated zone. If you add more water to the glass, the water infiltrates at the top of the ice cubes, percolates through the ice cubes in the unsaturated zone, reaches your water table, and recharges your aquifer.

What happens to your water table when you "rain" on your glass of water and recharge your aquifer? (tick-tock, tick-tock) It rises! This is an

Fig. 2.1 A glass of water as an aquifer

important difference between aquifers and oil reservoirs. Aquifers, like your phone, can be recharged whereas oil reservoirs are either not recharged or are recharged over millions and millions of years. Another important aspect of this simple experiment is that the size of the aquifer increased as the water table rose: there is now a greater ***saturated thickness*** or volume than there was before we recharged the aquifer in your glass. Saturated thickness is the vertical distance between the bottom of an aquifer (in this case, the bottom of your glass) and the top of the aquifer (in this case, the water level, or water table, in your glass).

Hopefully you have a (reusable) straw nearby and can put it into your glass. You can think of your straw as a *borehole* (a narrow shaft sunk into the subsurface). A borehole with a purpose—a ***well***—is used to extract oil, gas, or water. For our glass of water, we are going to use our borehole—the straw—to monitor our aquifer, turning the borehole into a *monitoring well.* If your straw is transparent or, like a boba straw, big enough to peer into, you should be able to see where the water is in the straw. How does the location of water in the straw compare to the location of water in your glass?

The water locations in the glass (aquifer) and the straw (monitoring well) are the same! This is what we refer to as a ***water level***, the level of water in an aquifer. We can measure from the top of the glass (land surface) to the water level (***depth to water***), which is 4 centimeters in my glass. Knowing the depth to water is important for, among other things, knowing how far water has to percolate to reach the water table (which tells us something about recharge), how deep wells need to be, and how far we have to lift water to get it out of the aquifer, something that affects the economics of producing groundwater. But it is also useful to compare the water level in one monitoring well to the water level in another monitoring well. In order to make this comparison, scientists often describe water levels as elevations relative to a benchmark.

For the benchmark for our glass-of-water aquifer, we will use the tabletop. The distance from my tabletop to the top of my glass (the land surface) is 10 centimeters; therefore, the land-surface elevation is 10 centimeters above my tabletop, or 10 centimeters AMT (above my tabletop). Since it was 4 centimeters from the top of my glass (land surface) to the water level in my well (straw), the ***water-level elevation*** is 6 centimeters AMT (the land surface elevation at the well minus the depth to water). Hydrogeologists also refer to a water-level elevation as a *hydraulic head*, but for this book, since the term is more descriptive, I will use water-level elevation. For real aquifers, water levels are typically benchmarked to above mean sea level (AMSL). Given that sea levels are rising, it is becoming more important to note which sea level you are referencing.

Water-level elevations are useful because they are a measure of *gravitational potential energy* where energy is measured by relative position. Just as a hunk of rock has a greater gravitational potential energy at a higher elevation above the Earth's (near) surface than one at a lower elevation, so too does water, especially since gravity plays such an important role in the movement of water. As the old adage goes, water flows downhill. More accurately, water flows down potential; this refinement is particularly important with groundwater because it commonly moves upwards in elevation. As we will see soon, water-level elevations are amazingly useful in telling us where groundwater came from and where it is going.

Let's return to our water molecule that has, thus far, changed from rainwater to infiltrating water to percolating water to recharging water and has now, finally, made it to the aquifer. Almost by definition, water has to move in an aquifer. If it doesn't, it will not have flushed out seawater from the original deposition of certain porous media or, if left to reside in porous media for thousands to millions of years, will become too salty (***saline***) for most uses. What controls how water moves through porous media?

Permeability

Henry Darcy was born in Dijon, France, in 1803, and attended college at L'École Polytechnique and L'Ecole des Ponts et Chaussée, both in Paris (Simmons 2003). While working for the Corps of Bridges and Roads, he designed a gravity-driven water supply and distribution system for Dijon that captured the flow of Rosoir Spring about 10 kilometers north-northwest of town. Late in his life, while he was ailing, he wrote a book, *Les Fontaines Publiques de la Ville de Dijon* (Darcy 1856; Bobeck 2006), that described his approach for designing Dijon's water system. In an appendix, he presents an equation, now known as Darcy's Law, that represents the steady flow of water through porous media, the experiments for which he conducted in the courtyard of the hospital he was recovering at.

Darcy's equation, which follows the same mathematical form that describes the flow of electricity and heat, relates the flow of groundwater to permeability, water-level elevations, distance, and cross-sectional area. ***Permeability*** refers to the ability of a porous media to transmit fluids (which includes liquids and gases). Porous media with a high permeability can more easily transmit fluids than porous media with a low permeability, all else being the same. Permeability is often used in a general sense (higher or lower) but can also be quantified through laboratory and field experiments where the term *intrinsic permeability* (intrinsic to the porous media and independent of fluid properties) is used. In the oil and gas industry, in honor of his contributions, the unit of permeability is the millidarcy. When water is the fluid and fluid properties are accounted for along with permeability, the result is the *hydraulic conductivity*: the ability of porous media to transmit water. In this book, since we will always be discussing groundwater flow, I will use the term permeability when referring to the general ability of porous media to transmit water, a common practice in the vernacular of the science.

You might think that the more pores you have in your porous media, the higher the permeability, but that is not necessarily the case. For example, the porosity of clay is about 40 to 70 percent while the porosity of gravel is about 25 to 40 percent (Freeze and Cherry 1979), yet water moves much more easily through gravel than clay. The reason for this is that the pores in clay, although numerous, are tiny and tortuous, making it much harder for water to flow through them than the much larger pores in gravel.

I briefly mentioned permeability when we were following our water molecule through the unsaturated zone. Percolating water is also influenced by permeability, but it is more complicated due to two-phase (air and water) flow and the presence of capillary forces. In fact, in the unsaturated zone, it can sometimes be easier for water to flow through a clay than gravel. Entire books, courses, degrees, and careers are dedicated to understanding water flow in the unsaturated zone (e.g., Stephens 1996[1]).

Groundwater Flow

I mentioned earlier that water levels are a measure of potential energy and that water moves from higher potential energy to lower potential energy. Let's say we have a one-dimensional groundwater flow system through a saturated, sand-packed pipe resting horizontally on a table. Both ends of the pipe are screened (to keep the sand in) and sealed but connected to vertical pipes filled with water, one for each end. The vertical, water-filled pipe on your right is 4 centimeters shorter than the one on your left. Let's also say that we have set up this experiment in a restaurant with excellent service such that the water-filled pipe on the left is always filled with water (Fig. 2.2).

Similar to our glass of water, we can measure the water-level elevation, relative to the tabletop, in each of the vertical, water-filled pipes. The left

[1] Dr. Stephens, despite my crazy goth hair and eyeliner, gave me my first job in hydrogeology as an undergraduate back in the mid-1980s.

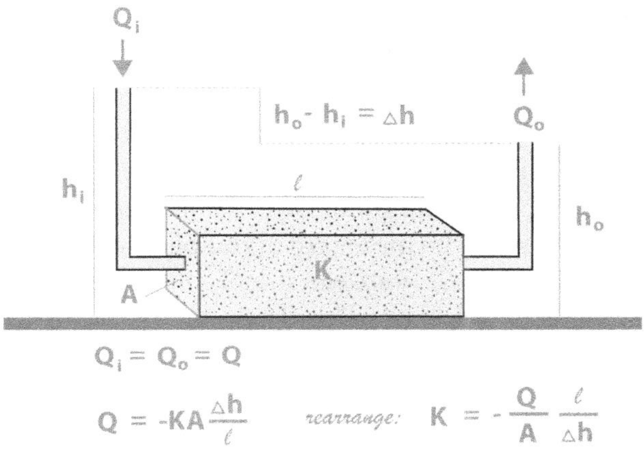

Fig. 2.2 Conceptual horizontal sand column experiment where Q_i is the flow into the column with units of length cubed per time, Q_o is the flow out of the column in units of length cubed per time, h_i is the water-level elevation (hydraulic head) on the input side of the column in units of length, h_o is the water-level elevation (hydraulic head) on the output side of the column in units of length, Δh is the difference between h_o and h_i, l is the length of the column, A is the cross-sectional area of the column in units of length squared, and K is the hydraulic conductivity in unites of length per time. The equation in the lower left is Darcy's Law. The equation on the lower right is Darcy's Law rearranged to solve for hydraulic conductivity, showing how hydraulic conductivity could be measured in the laboratory

side has a water-level elevation of 30 centimeters above the tabletop while the right side has one of 26 centimeters above the tabletop. Given these water-level measurements and the adage that water flows down potential, which way is the water flowing?

If you replied "left to right," you are correct, because the water-level elevation on the left is at a higher elevation (greater potential energy) than the water-level elevation on the right. Water generally flows from higher water-level elevations toward lower water-level elevations. I qualified the previous sentence with "generally" because, in the real world, there may be complications, such a *barrier fault*—a fracture with low permeability—where groundwater can't flow through easily. In our present case, we have a pipe filled with sand with no barriers, and the water flowing out of the pipe on the right side is creating a mess ("Waiter! More napkins!").

The difference between those two elevations is the change in the water level—the change in potential energy—across the length of the sand-filled pipe: that is, 4 centimeters. At this point, the length of the pipe becomes important. That driving force of 4 centimeters behaves differently in a pipe that is 30 centimeters long versus a pipe that is 15 centimeters long. Therefore, we are going to divide the difference between the two water-level elevations by the length of our pipe (which is 30 centimeters) to define the ***hydraulic gradient***: 0.133. If the pipe was half as long, the hydraulic gradient would be twice as large: 0.266. Note that a hydraulic gradient has no dimensions (as long as you use consistent units [as you should]) since it is defined as a length divided by a length.

The last term we need is a cross-sectional area which is, in our present case, the cross-sectional area of the interior of the sand-filled pipe. Given this, we can now calculate the flow, *Q*, through the sand-filled pipe with a cross-sectional area, *A*, with a defined hydraulic gradient, *i* (which we defined a short while ago [assuming you haven't fallen asleep yet] by the change in water-level elevation across a defined distance, the length of the pipe), and the hydraulic conductivity, *K*:

$$Q = -AKi$$

There are several things I want to point out about this equation: (1) this is the only equation I will present in the book, so don't post the book on ebay just yet; (2) this is the mathematical expression of Darcy's Law; (3) if we know the hydraulic conductivity, the hydraulic gradient, and the cross-sectional area, we can calculate how much water is flowing through the sand-filled pipe; (4) if we know the flow through the pipe, the hydraulic gradient, and the cross-sectional area, we can solve for K in the equation above and calculate hydraulic conductivity; and (5) we can use this equation in aquifers.

If you are wondering where that negative sign comes from, it's from the hydraulic gradient, which is technically negative because it's a negative slope. Even the purest of hydrogeologists do not report hydraulic gradients as negative values, but even the least pure hydrogeologist would have a fit if I left that negative sign off. Hydraulic conductivity has units of length per time, area has units of length squared, and hydraulic gradient is unitless resulting in volumetric flow of length cubed per time.

I debated whether or not to include a discussion of Darcy's Law in this book but decided to include it because it demonstrates how important the seemingly simple measurement of water levels is, providing us with an estimate of the direction of groundwater flow and critical information, combined with estimates of hydraulic conductivity, of how much groundwater is flowing. The equation above is for a one-dimensional flow system—there are more complicated equations for two as well as three dimensions. The equation also assumes *steady state*, or *equilibrium*, conditions where the flow and hydraulic gradient are constant in time. Accordingly, there are also *non-equilibrium* equations that describe changes in flow and water levels over time. The equation, whether in its form above or more complicated forms, also provides the mathematical foundation for numerical groundwater flow models. Cool.

Back to our water molecule!

Now that we understand the role of water levels and permeability, we can now discuss the movement of our water molecule in the aquifer. Ultimately, it's the water levels and permeability that decide what direction and how fast our water molecule moves. For our tabletop exercise, our sand-filled pipe represented one-dimensional flow. But in an aquifer, flow is two- or even three-dimensional. For the time being, we will focus on two-dimensional flow.

Ultimately, our water molecule is going to move in the direction of greatest hydraulic gradient following the path of least resistance. In real-world aquifers, hydrogeologists will measure water-level elevations in as many wells as they can so they map out a water-level surface, also referred to as a *potentiometric surface*. This is similar to measuring land-surface elevations and mapping contours to create a topographic map—but for groundwater levels. In many cases, the water-level surface is a subdued reflection of the land surface, in large part because aquifers tend to recharge under hilltops and discharge in valleys.

Just as a ball will tend to roll in the direction of steepest descent on a hill, groundwater will tend to flow in the direction of steepest downward hydraulic gradient. So after our water molecule recharges into the aquifer, it merges into an active groundwater flow system and moves in the direction of the steepest downward hydraulic gradient.

Thus far, we have assumed that our aquifer and its associated unsaturated zone are the same throughout, what is referred to as *homogeneous*. While some aquifers are, indeed, approximately homogeneous, many are not. Aquifers that are not the same throughout are referred to as *heterogeneous*. If our water molecule recharged into a sandy aquifer with pockets of clay, our molecule might have to divert around these clay lenses as it moved through the aquifer.

One last permeability concept to consider is *isotropy*. Isotropy means that the permeability of our porous media is the same regardless of which direction we flow water through it. For example, beach sand would be isotropic. Anisotropy means that the permeability of our porous media is different depending on which direction we flow water through it. For example, if our beach sand had long, horizontal, thin layers of clay in it, it would be more difficult for water to flow through the aquifer vertically than horizontally. Fractures in the aquifer may affect horizontal flow. In cases of anisotropy, groundwater may not flow in the direction of maximum hydraulic gradient because it also has to contend with local permeability perturbations.

There is a general saying in (near-surface) physics: What goes up must come down. For groundwater, I like to say: What goes down must come up.[2] Groundwater will generally follow a downward arcing path from areas of higher potential—generally beneath hills—to areas of lower potential—usually in valleys. Because of the arcing paths, it is not unusual for groundwater to flow upward. This is where pressure from areas of higher potential pushes the water upward toward a discharge point (Fig. 2.3). A fellow by the name of József Tóth was the first to publish about this phenomenon in 1963 (Tóth 1963).

If our water molecule recharged at the top of the hill, then it will flow from higher to lower water levels along the path of least resistance. In this case it will at first flow downwards and then begin to move laterally until, as it approaches the nearby valley the aquifer is hydraulically connected

[2] Yeah, yeah: I know that there are exceptions here, but please don't let academic elitism ruin a good euphemism.

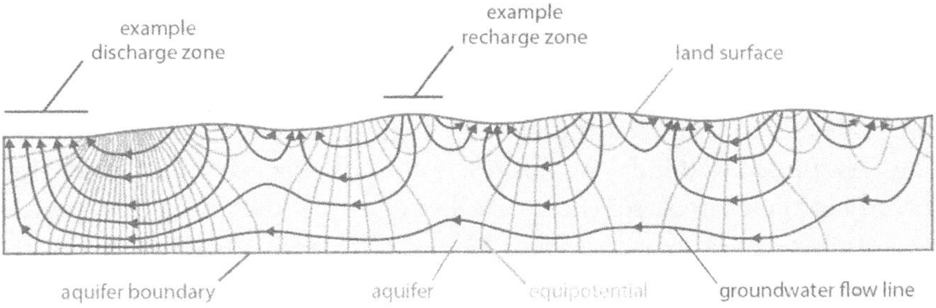

Fig. 2.3 Local, intermediate, and regional groundwater flow paths through an aquifer cross-section (after Freeze and Cherry 1979, p. 196). The top surface is the water table, which is a subdued representation of land-surface topography. The dashed lines are lines of equal potential energy, while the solid lines with arrows are groundwater flow paths indicating the direction of flow. Note how, at discharge points (where the lines end with upward arrows), groundwater flows upwards in elevation in the direction of downward potential

to, it begins to move upward and then discharge into a stream. If you are perplexed by water moving upwards, imagine if the flow path was a pipe of sand. On one end, just as in the restaurant, is the higher water level (or potential) and on the other is the lower water level (or potential). Just as in our horizontally one-dimensional pipe of sand, water flows from higher potential to lower potential, in this case upwards in the pipe connected to the left side of the pipe full of sand.

The time it takes for our water molecule to move through this *flowpath* depends on several factors including, from Darcy's Law, the length of the flow path, the hydraulic gradient, the hydraulic conductivity, and geometry of the aquifer (which helps define the cross-sectional area). In limestones with caves, it may take days or even hours for the water to travel along the full flowpath while in sandstones it may take decades to hundreds—and even thousands and tens of thousands—of years.

Valley discharge points can be local, where the arc of groundwater flow is from a hill to a local spring; subregional, where the arc of flow is from a hill to a stream a couple sub-basins away; or regional, where the flow is from a hill to a major river valley at a distance (Fig. 2.3). It all depends on the where the water is recharged and the geometry and properties of the aquifer.

Groundwater that discharges to the surface may provide water for local ecosystems and, if the flows are large enough and the surface-water system connected enough, along a river and into estuaries at the coast. Groundwater may discharge into more static surface-water bodies such as bogs, swamps, wetlands, and lakes. Groundwater may also discharge directly into estuaries and the ocean. For example, the interglacial period we are in today raised sea-levels 120 meters (Stanford and others 2011) putting the lower regions of many coastal aquifers under the sea. Many of these areas host subaqueous freshwater springs. Flow in a stream from an aquifer is called ***baseflow***.

Groundwater can also be discharged by ***phreatophytes***: plants that have roots that reach the water table and extend into the aquifer. Uptake of groundwater by phreatophytes can be so large that rivers and springs that flow during the winter months when the plants are dormant can go dry during the summer when the plants are active.

And there you have it: flow through an aquifer! There are additional complexities to consider, especially when producing water from an aquifer, which we will discuss in the next chapter.

Unconfined and Confined Aquifers

Hydraulically, there are unconfined and confined aquifers (Fig. 2.4). An ***unconfined aquifer*** is an aquifer bounded on the top by a water table. Accordingly, an unconfined aquifer is also referred to as a *water-table aquifer*. An unconfined aquifer is called unconfined because the aquifer is not capped by a ***confining layer***, which is generally an aquitard. In an unconfined aquifer, the water level in a well will rise to the level of the water table. The aquifer we have been discussing with our water molecule is an unconfined aquifer as was our glass of water. The thickness of an unconfined aquifer can change over time depending on the amount of recharge (as in our glass of water), groundwater flow out of an area, or, as we will discuss later, groundwater production.

A ***confined aquifer*** is an aquifer that is bounded at the top by a confining layer such that the water in the aquifer is under pressure. Note that in both unconfined and confined aquifers, there is a confining layer at the

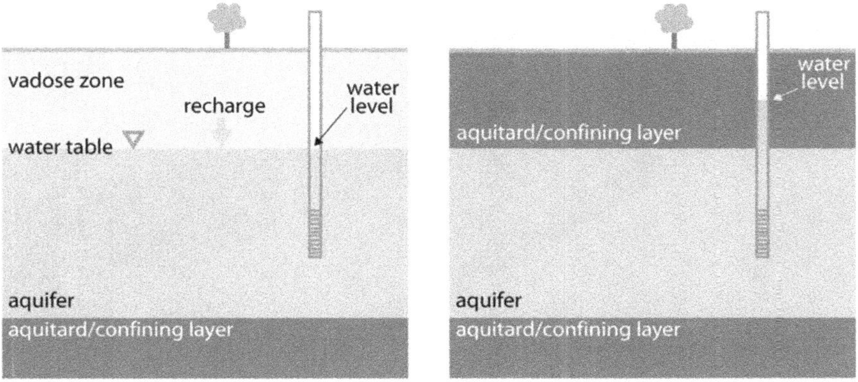

Fig. 2.4 Elements of unconfined (left) and confined (right) aquifers

bottom of the aquifers (the bottom of the glass in our glass of water). In a confined aquifer, the water level in a well will rise above the top of the aquifer because the water is under pressure. Sometimes, with the proper topography and hydraulics, the water level in a well will rise above the land surface—as I described in Chap. 1—and flow, what is referred to as a ***flowing well***.

A confined aquifer is also known as an ***artesian aquifer***. Note that producing flowing wells is not required for a confined aquifer to be considered artesian—the only requirement is for water pressure to push water levels in the aquifer above the top of an aquifer that is capped with a confining layer.

Almost all aquifers that have a confined portion also have an unconfined portion (Fig. 2.5). After all, an aquifer that is confined generally needs a part that is unconfined to receive recharge. Many confined aquifers have *dips*—that is, the geologic layers that host the aquifers have a slope to them. The up-dip part of the aquifer, which is at a higher elevation, is where the unconfined portion is while the down-dip part is where the confined portion is. Sometimes hydrogeologists talk about a *transition zone* where, depending on the water-level elevation, the aquifer may be unconfined in the summer (the water level is below the confining layer and thus a water table) but confined in the winter (the water level is above the bottom of the confining layer). To be complete, there are also

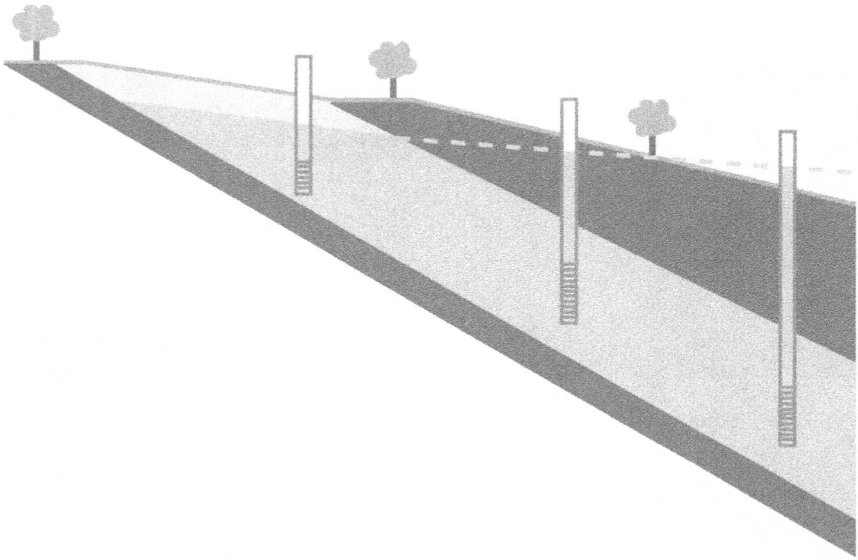

Fig. 2.5 Cross-section of a dipping aquifer showing unconfined and confined portions

semi-confined aquifers: aquifers that are confined but the confining layer is leaky, that is, can be a source of water to the aquifer when water levels drop in the aquifer.

Whether an aquifer is confined or unconfined is important, in part, for water quality, especially related to human activities on the land surface (it is far easier to pollute an unconfined aquifer than a confined one) and for how aquifers respond to pumping, which we will discuss in the next chapter.

Geology of Aquifers

As mentioned at the beginning of this chapter, what is or isn't an aquifer depends on who is using it. Although clays and shales, which have low permeabilities, are not traditionally considered aquifers, shallow wells in the near-surface weathered zone where erosional unloading, freeze-thaw cycles, and plant roots may have fractured the geologic media may provide small amounts of water, enough to meet the modest needs of a frontier homestead or a current, single-family home. More typically, aquifers consist of sands, gravels, sandstones, and limestones.

Sands and gravels are typically *alluvial*—deposited by rivers and streams—but can also be windblown (*aeolian*) and accumulated along shorelines. Sands and gravels are referred to as *unconsolidated deposits*—geologic media that has not been naturally cemented into rock. As you might imagine, alluvial deposits are typically located along streams and rivers; however, these deposits can be massive. For example, the Ogallala Aquifer in the Great Plains of the United States covers about 450,000 square kilometers and mostly represents the erosion of the eastern face of the Rocky Mountains. Sands and gravels, in large part because of their large *grain size* (the size of the sand grains and the gravel pieces) and lack of *cementation* (cement fills pore space), have high permeabilities; therefore, they can produce large amounts of water.

Sandstones are generally formed from ancient beaches as sea-level rose and fell. As the name suggests, sandstones are *consolidated deposits*—geologic media that has been naturally cemented into rock, generally over a long period of time. Sandstones represent the erosional bits of long-lost mountain ranges when the continents were arranged differently, so knowledge of local and regional geology is key in locating them. Sandstones, because they are cemented, do not generally have as high a permeability as unconsolidated sands, but do tend to have high enough permeability to produce good wells for a variety of uses.

Limestones were generally created in shallow, off-shore seas and are the result of precipitating calcite as well as the collection of the calcareous shells and debris of various critters that created a pelagic mud that consolidated into rock over time. Limestones, in and of themselves, are not particularly permeable. However, once exposed to the land surface and the acidity of percolating water, they often dissolve, creating large voids—sometimes caves—through which large amounts of water can flow. Limestones that are partially dissolved in this manner are referred to as *karst*, a word first used in science by Johann Weikhard von Valvasor in 1689 when describing underground rivers at Lake Cerknica (Larsen, 2003).

Another category are aquifers that achieve their permeability through fractures. A *fracture* represents a break in geologic media such that a rock is broken into two or more pieces. These fractures can dramatically improve the permeability of a rock because it is generally easier for water to flow through them. One example are fractured granites. Granite, in and of itself, has little to no permeability. But if that granite is fractured, it can transmit sufficient amounts of water to be considered an aquifer.

Broadly, there are two types of fractures: joints and faults. A *joint* is a fracture with no displacement along the fracture plane. It is simply a crack in the rock. Earlier, we discussed how shale, generally a confining layer, can be an aquifer near the land surface due to weathering. Weathering often leads to many joints in rocks near the land surface from the release of overburden pressures (rocks that used to compress the area now gone due to erosion), freeze-thaw action, or root activity that can pry small fractures into larger ones. A *fault* is a fracture with displacement, in other words, the porous media on either or both sides of the fracture has moved. Faults can be large, like the infamous San Andreas Fault in California where one continental plate is moving northward while another is moving southward, or be used to describe a single fracture. Faulting tends to create swarms of fractures, thus greatly increasing permeability and often facilitating the creation of karst aquifers by allowing acidic waters to more deeply penetrate and dissolve limestones.

At the End of the Flow System

In summary, an aquifer broadly refers to geologic media, such as rocks and sediments, that bears and transmits usable amounts of water. An aquifer is generally in the wells of the beholder: a city's aquitard may be a household's aquifer.

The groundwater cycle begins with the infiltration of water into the subsurface. That which isn't captured by plants or evaporated from the land surface percolates through the unsaturated zone and recharges the aquifer through the water table. Based on water levels (hydraulic heads) and permeability of the aquifer material, groundwater flows down gradient (down potential) towards discharge points, generally springs and seeps before the advent of water wells.

Depending on the geology, an aquifer may be unconfined (have an unsaturated zone and a water table) or confined (fully saturated and under pressure such that water rises above the top of the aquifer, also called artesian). Aquifers come in many shapes and forms but are commonly made from the sands and gravels of alluvium, sandstones, limestones (especially if they are karstified, i.e., partially dissolved), and fractured rocks.

Questions

What is the name of an aquifer in your area?
What is the aquifer made of?
Where is the aquifer recharged? And where has it historically naturally discharged?
Is this aquifer unconfined or confined in your area?
Are water levels measured in your aquifer? What do water-level elevations say about which direction groundwater is flowing?

References

Arago, F., 1834, Sur les puits forés, connus sous le nom de puits artésiens, des fontaines artésiennes, ou de fontaines jaillissants: Bureau des Longitudes, Annuaire pour l'an 1835, Paris, p 181–258.

Bobeck, P., 2006, Henry Darcy in his own words: Hydrogeology Journal, v 14, p 998–1004.

Byrne, D., Eno, B., Frantz, C., Harrison, J., and Weymouth, T., 1980, Once in a Lifetime: *in* Remain in Light, Track 1, Side 2, Sire Records.

Canadell, J., Jackson. R.B., Ehleringer, J.B., Mooney, H.A., Sala, O.E., and Schulze, E.D., 1996, Maximum rooting depth of vegetation types at the global scale: Oecologia, v 108, n 4, p 583–595.

Darcy, H., 1856, *Les Fontaines Publiques de la Ville de Dijon*.

Fetter, C.W., 2001, *Applied Hydrogeology* (4th Edition): Prentice-Hall, 598 p.

Foxx, T.S., Tierney, G.D., and Williams, J.M., 1984, Rooting depths of plants relative to biological and environmental factors: Los Alamos National Laboratory report, 26 p.

Freeze, R.A., and Cherry, J.A., 1979, *Groundwater*: Prentice-Hall, Inc., 604 p.

Larsen, P., 2003, Scientific Accounts of a Vanishing Lake–Janez Valvasor, Lake Cerknica, and the New Philosophy: Master's Thesis, MSc in History of Science and Technology, Oxford University, 52 p.

Lohman, S.W., 1972, Ground-water hydraulics: U.S. Geological Survey Professional Paper 708, 70 p.

Mace, R.E., Ridgeway, C., and Sharp, J.M., Jr., 2004, Groundwater is no longer secret and occult—A historical and hydrogeologic analysis of the East case: in Mullican, W. F., III, and Schwartz, S., editors, 100 Years of Rule of Capture: From East to Groundwater Management, Texas Water Development Board Report 361, p 63–88.

Meinzer, O.E., 1923, Outline of ground-water hydrology with definitions: U.S. Geological Survey Paper 494, 71 p.

Merriam-Webster, 2021, Permeable: Merriam-Webster.com Dictionary, https://www.merriam-webster.com/dictionary/permeable, accessed 23 Mar. 2021.

Norton, W.H., 1897, Artesian wells of Iowa: Iowa Geological Survey, v 6, 130 p.

Schenk, H.J., and Jackson, R.B., 2002, Rooting depths, lateral root spreads and below-ground/above-ground allometries of plants in water-limited ecosystems: Journal of Ecology, v 90, p. 480–494.

Simmons, C.T., 2003, Happy 200th birthday Mr Darcy and our thanks for your law! A tribute editorial celebrating the life and times of the father of our science, Henry Darcy (1803–1858): Hydrogeology Journal, v 11, p. 611–614.

Stanford, J.D., Hemingway, R., Rohling, E.J., Challenor, P.G., Medina-Elizalde, M., and Lester, A.J., 2011, Sea-level probability for the last deglaciation – A statistical analysis of far-field records: Global and Planetary Change, v. 79, n. 3, p. 193–203. https://doi-org.libproxy.txstate.edu/10.1016/j.gloplacha.2010.11.002

Stephens, D.B., 1996, *Vadose Zone Hydrology*: CRC Lewis Publishers, 347 p.

Tóth, J., 1963, A theoretical analysis of groundwater flow in small drainage basins: Journal of Geophysical Research, v. 68, n. 16. P. 4795–4812.

3

Domesticating Groundwater

"What makes the desert beautiful is that somewhere it hides a well."
Antoine de Saint-Exupery (1943)

Human use of groundwater started with simply knowing where springs were located so they could be visited to quench thirst. This was particularly important in drier climates where rainfall and sources of surface water are scarce. Then, with time, humans began to enhance springs or chase the water into the subsurface through digging a hole—a well! As sinking wells became more advanced, deeper and deeper aquifers could be tapped. Production technologies also advanced with time, with human power changing to animal power to steam power to electric and gas pumps. Each advance—and the increase in human population and the concomitant increase in the need for water—increased the use of groundwater, threatening the long-term viability of the world's aquifers. In this chapter, we trace the history of groundwater production to modern times and discuss the consequences of producing too much from an aquifer.

R. E. Mace, *Groundwater Sustainability*, Palgrave Studies in Environmental Sustainability, https://doi.org/10.1007/978-3-031-13516-3_3

Enhancing Springs

As human numbers and technological know-how increased, people began to modify springs to increase their flow by removing obstructions (such as rocks and debris) as well as direct their flow to irrigate crops. Archeologists sometimes refer to controlling and producing water as the domestication of water (for example, Garfinkel and others 2006). It is unclear exactly when springs were first tapped for use by humans, but it is not surprising that there's evidence of Neanderthals at spring sites, sites that also served to preserve that evidence for later uncovering (Clark 1944). It is also easy to imagine early humans seeking to protect and enhance springs by building borders around them to preserve water quality and improve access (Clark 1944).

At some point, someone figured out that they could direct springflow through hand hewn channels to bring spring waters to a more favorable camping or collection spot. Eventually, that turned into the more elaborate engineering of canals. Knossos and Tylissos in Crete used aqueducts during the Bronze Age (3300-1100 BC) to move springflows to their locations (De Feo and others 2013). The Nabataean cities of Petra and Humeima in present-day Jordan were among a number that were supplied by small, spring-fed aqueducts (De Feo and others 2013). Perhaps most famously, there are the Roman aqueducts, many of which (particularly for Carthage in Tunisia, Nîmes and Fréjus in France, and Side and Patara in Turkey) are fed by springs (De Feo and others 2013). Henry Darcy, discussed in the previous chapter, captured springflow and used gravity to supply water to hospitals and fountains in Dijon (Fig. 3.1). Some communities continue to tap directly into springs for their needs, often containing their spring in a concrete enclosure, such as Darcy did, to prevent an animal—or humans—from fouling the source. The City of Del Rio in Texas runs pipes directly into a spring orifice to capture the water just as it emerges from the ground.

Fig. 3.1 La Fountaine de Jouvence d'Hier a Aujourd'hui north-northwest of Dijon, France (photo taken by the author in 2016). Henry Darcy captured the flow from this spring and designed a gravity-fed aqueduct to carry the water to the City of Dijon 11 kilometers to the south-southeast. In developing the project, Darcy sought to preserve flows in the nearby creek for downstream users and took care of the needs of the communities between the spring and Dijon. The springhouse protects the spring from animals and humans, thus protecting water quality, and the overflow allows the spring to continue to contribute flows the creek

Sinking Wells

In archaic times, people referred to springs as wells. The word comes from the Middle English "welle" which is associated with the Old English "weallan," which means to bubble and boil (Merriam-Webster 2021). Therefore, you will sometimes see springs referred to as wells, such as the Turnbridge Wells in England, named for the iron-rich springs that flowed in the area. At some point, the word "well" was attached to human-dug shafts to access groundwater.

We don't know when the first well was sunk, but it probably happened at an ephemeral spring that, as the flow declined and perhaps stopped, someone dug into the ground at the spring location to continue to access water. With time, as water levels fluctuated, this digging probably turned into something of a shaft, what we would recognize today as a hand-dug well. Or perhaps it happened in a stream bed under similar circumstances. Animals such as wolves have been observed to dig in a dry stream bed in search of water (Meinzer 1934). Watching humans may have taken note.

After realizing that one could chase and find water digging at a spring, someone made the leap of faith that digging a shaft near the spring—not just *in* the spring—could also provide water. This probably also happened gradually, with someone sinking a well close to the spring before slowly moving out from the spring. Perhaps the motivation was getting the well out of a drainage channel where floods might fill the borehole with sediment. Finally, someone—the first wildcatter—realized that water in the ground was relatively common and one only had to sink a well to find it.

The earliest known water wells are from about 8000 BC in Cypress (Peltenburg and others 2001). These wells, about 2 meters in diameter—big enough to accommodate one or two well diggers—and up to 8 meters deep, were hewn from solid rock (Streit and Garfinkel 2015). In looser sediments, the sides of the well would have needed shoring, what's known as *well casing*, to prevent the walls from collapsing. Water wells dug during the Neolithic with carefully made wooden casings have been found in Central Europe with the oldest dating to 5196 BC (Rybníček and others 2018). Neolithic well sinkers dug these wells using antlers, bones, and specialized wooden tools (Weiner 1994) to dig as deep as 14 meters (Rybníček and others 2018) to reach the water table.

Well digging slowly progressed over time, including using stone instead of wood for casing. Particularly impressive is the ancient city of Mohenjo-Daro (2450 BC) in Pakistan which relied on a projected network of 700 wells, each several meters deep, to meet local water demand (Jansen 1989). Vitruvius (~75 to 15 BC) described in his set of books titled *De Architectura* how to locate promising places to dig wells and how to safely dig them.

Building Qanats

The arguable apex of hand-dug technology was reached in Persia with qanats, the subsurface fusion of well digging and aqueducts. A *qanat* is a gently sloping tunnel (the first horizontal well!) with vertical shafts that intersects the water table at a higher elevation—generally in alluvium along a mountain front—that conveys water by gravity to a lower elevation where the qanat becomes exposed at land surface (Fig. 3.2). The uppermost vertical shaft is known as the mother well, which also intersects the water table (Goes and others 2017). Subsequently dug shafts assisted in accessing the tunnel for digging and maintenance (and perhaps water).

Probably invented in or near Iran (Wilkinson and others 2012), qanats are common in the Middle East but have been constructed as far afield as western China, North Africa, and Spain (Taghavi-Jeloudar and others

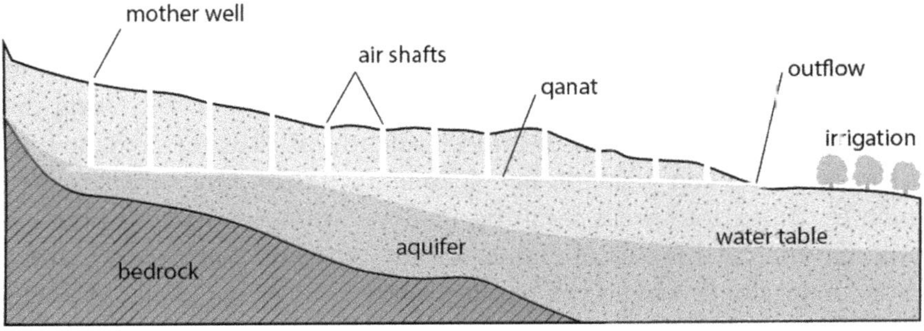

Fig. 3.2 Cross-section of a qanat (after Goes and others 2017)

2013) and even in the Central Andes of modern-day Peru, where they are named puquios (Lane 2017). Qanats were first constructed about 2000 BC (Fattahi 2015) in Iran and often took decades to build due to great depths (upwards of 300 meters) and great lengths (the qanat for Zarch, Iran, was constructed 3000 years ago, continues to operate today, and is 71 kilometers long). Planning and constructing qanats demonstrates a clear understanding by proto-hydrogeologists of how local aquifers worked, perhaps inspired by ancient copper mines that released water as a side-effect of mining operations (Taghavi-Jeloudar and others 2013).

Drilling Wells

The Chinese are credited with advancing methods for drilling deep wells as early as 256 B.C. (Hallenburg 1997, p. 17). They developed the churn method of drilling a well—often referred to now as the cable-tool or percussion method—and introduced the concepts of well casing to keep loose soil and sand from entering the borehole and drilling muds to maintain borehole integrity and carry well cuttings to the surface. The churn method involves slowly chipping away at the bottom of a borehole with a sharp bit attached to the end of a string of poles. The bit chips at the bottom of the hole by repeatedly raising and lowering the string of poles—and hence the bit—in the borehole. By 300 A.D., the Chinese were drilling wells up to 140 meters deep, generally in the pursuit of brackish groundwater to evaporate for its salt (Jung and Fang 1948). By 1700, the Chinese in the district of Wu-Tung-Kiao had drilled over 10,000 wells up to 1242 meters in depth.

The earliest record of deep well drilling in Western Europe comes from Carthusian monks at Sillers near Lille, France, in the 1100s. Using the churn method, monks in 1126 drilled to depths of 100 meters to tap into a source of flowing groundwater. Monks in nearby Gonnehem subsequently drilled four wells resulting in flows strong enough to power a mill. Given that this part of France was known as the Province of Artois, these magically flowing wells became known as artesian wells (Tolman 1937, Villiers 2003).

In 1828, a French missionary in China by the name of Abbe Imbert sent letters home describing in detail the Chinese method of drilling deep wells. A French engineer named Jobard was the first to put the method to use in Europe. Within 15 years, churn drilling was common on the European continent (Conner 2005). However, it was a deep well drilled near Paris, France, that captured the imagination of the western world on the possibilities of deep artesian wells.

The well, drilled by a self-taught engineer named Georges Mulot, was a 20.3-centimeter diameter hole drilled in Grenelle, a neighborhood in modern-day Paris. It was with this well, started in 1833, that western drillers learned—over the course of eight years—how to drill deep into the earth. A key part of this effort was developing the methods and tools to recover broken well strings in the borehole. At one point, the Grenelle crew dropped 82 meters of drilling rods into the bottom of the borehole. It took the ingenuity of Mulot and 15 months of patient downhole fishing to retrieve the pieces before drilling could resume.

At a cost of 400,000 francs, Mulot's team reached a depth of 548 meters on February 26, 1841, where they hit a cavity that produced pure water that rose 36 meters above the land surface and flowed at 35 liters per second (Mackworth 1854, Swindell and Burnell 1883, Cox 1895) and gushed 33 meters into the air (Meunier 1978; Fig. 3.3). A subsequent book published in 1847 by the French engineer Joseph Degousée—recognized worldwide at the time as an artesian well-drilling expert—helped to further disperse knowledge on how to drill for artesian water (Degousée 1847).

Artesian wells were seen as particularly useful and valuable if they produced ample amounts of fresh water. River water and shallow well water, especially that with exposure to sewage, were classified as "dangerous" while water from springs, upland surface water, and deep well water were classified as "wholesome" by water resource authorities at the time (Bell 1881).

The Grenelle Well inspired many across the world to pursue deep artesian waters of their own. This well and other subsequent French gushers, inspired the drilling of the first deep artesian wells in the United States located in Charleston, South Carolina, and Louisville, Kentucky.

Fig. 3.3 Engraving of the well in Grenelle, France, after attaining artesian flows as recorded by Louis Leroy (scan of an original print from the author's collection)

Charleston experimented several times with deep artesian wells. The city started its first well in 1824 and reached a depth of 102 meters before it had to be abandoned due to the drilling rods twisting off in the hole. Private interests attempted a second well in 1826 but ran out of money before the well could be completed. The United States government attempted a third well at Fort Sumter in 1846 but stopped at 110 meters after the iron rods twisted off and appropriations ran out. The city attempted another well in 1847 but stopped at 85 meters. In 1848, the city hired Mr. A.C. Welton, a driller who had had drilling success in Alabama and other nearby states, to drill a well that ultimately reached a depth of 384 meters by 1854 or 1855 and produced a small flow of 2 liters per second (Wells and others 1854, Whitney 1894, Lynch and others 1884).

William Belcher, in his search for better quality water for his sugar refinery in St. Louis, commissioned a deep well in search of artesian water (American Journal of Science 1853). His crew began drilling a borehole in the spring of 1849 at Second and O'Fallon streets using hand power. After 18 months of drilling, the borehole had only reached a depth of 67 meters, after which Mr. Belcher commissioned the use of a steam-powered drilling apparatus. With this new technology, his drilling team reached a depth of 670 meters by March 1854 (Litton 1860) at which point the well produced highly mineralized water at 4.7 liters per second (Shufeldt 1867, p. 20). The Belcher well struck gas and brackish water but ultimately did not tap a fresh flowing water source. Even though Belcher plumbed most of the salty "sulphuretted" flow to the city sewer, the well's water found its fans for its medicinal and remedial virtues (Litton 1860, Bullard 2004). The water was later used for the Belcher Water Bath Company, whose bathhouse operated until the 1970s (Bullard 2004).

The relative success of these wells led to others to drill test holes in search of deeper water, including Mr. Patten in Fort Worth described in Chap. 1.

Elevating Water

If groundwater did not flow of its own volition from the subsurface, then people had to somehow produce the water from the borehole. If the depth to water wasn't too great, people could fashion a scoop or ladle out of a gourd, animal skin, or pottery to pull water from the well. Sometimes societies built infrastructure to ease access to groundwater. The stepwells of India are particularly impressive for their massive scale and, consequently, their ability to service multitudes of people seeking water. As the name suggests, step wells are wells with steps that descend to the water, the first ones built in 550 AD (Voudouris and others 2019, Fig. 3.4).

Because water is heavy and can be deep—and the hard labor of fetching water often fell to women—people began using machines to simplify access to it. The *shaduf* is the first machine used to access groundwater in wells in several ancient civilizations (Yannopoulus and others 2015). A shaduf is a leveraged pole with a short rope and a water-holding receptacle on the end, first used in the Middle East and India about 3000 BC (Chondros 2010). Shadufs are called well poles or well sweeps in the West. They can produce about 2.5 cubic meters per day (Yannopoulus and others 2015), enough to water 0.1 hectares of land in 12 hours (Mays 2010).

Fig. 3.4 The Chand Bawri step-well in Abhaneri, Rajasthan, India, built during the eighth and ninth centuries (photo by Julie Alexander)

Deeper wells required a rope and human effort to raise water. At Mohenjo-Daro, ruts in the sides of the wells indicate the use of ropes to pull water up from the depths (Clark 1944). Because this was labor-intensive and limited production amounts, the Romans invented the *waterwheel* about 600 to 700 BC, often referred to as the Egyptian waterwheel, or *noria* (Yannopoulus and others 2015). A waterwheel is a horizontal axis wheel powered by water flow that also captures water in a series of rotating buckets attached to the wheel that dump the water into a trough for use (Yannopoulus and others 2015). Without the aid of flowing water, the waterwheel could be operated by animals such as camels, oxen, mules, and horses. Persian waterwheels, which were equipped with two pulleys to facilitate animal power, could produce 10 cubic meters per hour for a 9-meter water lift or 20 cubic meters per hour for a 1.5-meter lift (Yannopoulus and others 2015). The waterwheel was competitive with steam engines well into the Industrial Revolution (Yannopoulus and others 2015). A later improvement to the water wheel, to retrieve water from greater depths, was the *tympanon*, which used a chain of buckets instead of buckets attached to the wheel (Yannopoulus and others 2015).

Ktesibios of Alexandria was the first to use pistons to pump water, developing a dual piston suction pump to lift water from wells in 250 BC (Vitruvius n.d.). Romans used this technology with bronze pistons with leather seals bored into wood about 300 AD to pump water from a well 16 meters deep in Vicus, Belgium (Deutsches Museum 2021). The technology was lost during the Dark Ages but was reintroduced by Georgius Agricola in *De re metallica* published in 1556 for mine dewatering as well as pumping water wells (Agricola 1556). These pumps create a suction in a sealed cylinder to pull water into them and then uses a valve system to move the water upward and onward.

Because suction pumps can only pull water from about 10 meters deep before breaking suction, the pump apparatus had to be placed within that distance of the water level (and ideally less to increase the efficiency of the pump). If the water was deeper than 10 meters, then operators had to dig a pit to get closer to the water. Suction pumps were originally human-operated before people looked to the wind to capture groundwater.

Wind for Water

The first documented evidence of capturing wind to raise water comes from Sistan in eastern Persia in the tenth century where panemones—fixed, vertically oriented windmills with blades on a shaft—were used (Shepherd 1990, Fig. 3.5) and are still used today. In the 1100s, horizontal axis windmills began to appear in Europe while references to windmills began to appear in Chinese texts in the 1200s (Shepperd 1990). The first European-style windmills were fixed with respect to the wind, so they were oriented toward the wind where the rate of rotation was controlled by the sails on the blades (Hills 1994[1]). A later improvement included placing the body of the windmill on an enormous post such

Fig. 3.5 Panemone windmills of Nashtifan in Khaf, Khorasan Razavi, Iran (photo by Ahmad Hassanzadeh). These windmills date to over a 1000 years ago

[1] The rest of the paragraph is based on this reference.

that the face of the windmill could be manually turned into the wind. A subsequent improvement, called a wingtail, could automatically turn the face of the windmill into the wind.

European-style windmills were large and unwieldly and, even with a fantail, required considerable labor to maximize power production or shut the windmill down in storms to protect it from damage. In the 1840s, John Burnham, who co-owned a pump factory in Connecticut, had the idea of developing a small, self-regulating windmill to power his company's pumps (Baker 1985; Hills 1994[2]). In 1854, he invited Daniel Halladay to design such a device, one that would automatically orient itself into the wind and automatically spill the wind to maintain a manageable number of revolutions per minute. As a result. Halladay developed the world's first self-regulating windmill and started the industry for what became known as the American-style windmill. Because these windmills were smaller in size, could be mass-produced, and required less capital outlay, they became the preferred tool to produce water as American railroads and settlers moved westward.

Improvements in American windmills happened quickly as numerous companies sprang up in northern Illinois to meet demand. A notable improvement came because of Thomas O. Perry's experiments with metal windmill blades in 1882. Perry worked for the U.S. Wind Engine and Pump Company, which Burnam originally formed in Chicago to sell Halladay windmills to westerners and eventually eclipsed the parent company to become the largest windmill manufacturer in the world. Perry's supervisors charged him with assessing metal windmill blades to determine if they were superior to the Company's wooden blades and therefore a threat to their business model.

In response, Perry developed a whirling table to test the efficiency of different windmill heads and blade designs. After seeing the promise of metal blades, he then perfected the number and shape of the blades through more than 5000 trial-and-error tests to develop a windmill that was 87 times more efficient than the best wooden windmills of the time. His employers did not adopt his design, deciding to stay with wooded blades.

[2] The rest of this section is based on these two references unless otherwise specified.

Consequently, Perry left the U.S. Wind Engine and Pump Company and teamed up with Alfred Noyes to form Aermotor, whose windmills were derisively called "mathematical windmills" by its competitors. The company only sold 24 windmills in 1888 but was selling 20,000 a year by 1892, and, by 1900, Aermotor had half the American windmill market. Metal windmills claimed more than three-quarters of the market by 1914 with only one wooden windmill company remaining in 1940. After a stint in Arkansas and Argentina, Aermotor still exists today, churning windmills out of its manufacturing plant in San Angelo, Texas.

American-style windmills convert rotary motion of the windmill head into an up-and-down motion (the following description is based on Calderone 2018). This up-and -down motion then drives the pump, or sucker, rod up and down in the well and in a cylinder with a check-valved plunger that pulls water into the pipe with a downstroke and up the pipe with an upstroke.

Because wind is often intermittent, there is often a storage tank nearby to hold water for use when the wind fails. Excess water is generally discharged to the land surface. How much that can be pumped depends on the depth to water, the size of the windmill, the size of the cylinder, and the wind. A standard windmill with a 2.4-meter diameter wheel, a 4.4-centimeter cylinder, a depth to water of 56 meters, and 24 to 32 kilometers per hour winds can pump about 0.16 liters per second. That's enough for a farmstead and a garden but not enough for irrigation.

Centrifugal Forces

Denis Papin (1647 to 1713), a French physicist and inventor, developed the first centrifugal pump in 1689 where water that enters the center of the pump is flung by centrifugal force to the edge of and out of the pump casing by a vaned impeller (Papin 1689, Harris 1953,[3] van Esch 1997). le Demour (1732) developed a crude pump to elevate water, reaching about 2 meters in height. In 1818, the "Massachusetts Pump" was invented (the inventor is lost to history) and introduced the forced vortex, a technical

[3] Information in the rest of this paragraph sources from this reference unless otherwise noted.

development still used today. In 1839, the United States granted W.D. Andrews a patent for a pump similar to the Massachusetts Pump. Combes (1838, 1840) presented a paper investigating curved blades in a centrifugal pump, a critical advancement. John George Appold, a British engineer, empirically designed a curved vane centrifugal pump in 1849, a pump that was more than three times as efficient (at 68 percent) than existing pumps, earning him a medal at the Great Exhibition in the Crystal Palace. By 1900, centrifugal pumps were common across Europe and the United States with larger units capable of producing upwards of 20 liters per second.

These days, centrifugal pumps are also known as suction pumps or trash pumps. They earned the latter name because they can pump some pretty nasty water, including water laden with silt and (soggy) trash. They are also known as suction pumps because, in drainage operations—as well as early well pumping uses—the pumps pull water into them through a pipe or hose placed into the water and need priming to start the pump. Theoretically, a suction pump can pull water up from 10-meters deep, but, in the real world, 6 meters is a more realistic depth. If the water table was more than 6 meters below land surface, then a pit would have to be dug to bring the pump closer to the water table. For example, if the water table was 12 meters below land surface, you would have to dig at least a 6-meter-deep pit (probably more) to get the pump close enough to the water table to pump. Early centrifugal pumps were powered by steam engines through belts.

In the United States, a centrifugal pump was used for irrigation as early as 1885 near Eaton, Colorado (Steinel 1926). In 1896, farmers near Garden City, Nebraska, pumped the Ogallala Aquifer with centrifugal pumps (Green 1973[4]). However, these pumps and pits were expensive and, in many places, the depth to the water table was too deep. What was needed was a deep-well pump that didn't require a pit. Preston K. Wood of Los Angeles, California, was the first to design such a pump in 1897 called a turbine pump, a type of centrifugal pump. However, Wood's pump was not very efficient. In 1901, Byron Jackson of San Francisco,

[4] The rest of this section is based on this reference unless otherwise noted.

California, developed another deep-well turbine pump design for the Pabst Brewing Company of Milwaukee, Wisconsin.

Rice provided the second-to-last boost to the adoption of turbine pumps in groundwater production. An increase in United States rice tariffs in 1897 created a rice boom along the Gulf Coast, including Texas. With limited surface water and frequent droughts, rice farmers needed to produce groundwater to maintain production. In 1901, Mahlon E. Layne moved to Texas after visiting the oil well field in Spindletop convinced that the future of his business was in servicing oil well drilling. He hailed from South Dakota drilling large-bore artesian wells finished with well screen of his own design. After successfully deploying his well screens to facilitate oil production, he began deploying them for rice irrigation wells in the area. One day in 1902, after climbing out of a well pit, he came up with the idea of a pit-less pump—a pump that would not require a pit. He designed a pump and, after a series of modifications, patented it in 1906. In 1907 he formed the Layne and Bowler Company to build and market his pumps, which—aquifer willing—could produce 1000 gallons per minute.

Affordable power was the last needed boost. Electricity was initially the preferred power for deep-hole turbine pumps, but electricity in those early days was expensive and not available everywhere. Green (1973) noted that in 1910, running a pump on electricity cost more than running a pump with a steam engine and a four-person crew. Ultimately, it was the internal combustion engine that finally brought an affordable power plant to the wellhead, first with simple oil-fueled engines and then with Detroit V8s, as well as the electrification of rural areas (Fig. 3.6). As the pumps got smaller and more efficient and the power plants also got smaller and more efficient, groundwater production began to increase dramatically across the United States and the world.

Groundwater in Storage

As we saw in Chap. 1, one of the attributes of groundwater is the great amount of water in storage. Just as we have unconfined and confined aquifers, we also have ***unconfined and confined storage***, which fall under

Fig 3.6 Centrifugal pump installed to pump water sourced from San Solomon Springs near Balmorhea, Texas, powered by the engine from a buried truck (photo by the author)

the uniform term of *storativity* or *storage coefficient*. Many technical terms used in the study and evaluation of groundwater resources have a development bias because the goal, in the early days, was to tap these resources for human and economic purposes. Storage seems to have that bias, so I will use the more unwieldy storativity, *unconfined storativity*, and *confined storativity* for the rest of this discussion.

Storativity refers to how much water is released from an aquifer, whether to a well, a spring, or to the flow system. In an unconfined (water-table) aquifer, this water sources from draining part of the aquifer. If you think back to our glass of water with the straw, imagine taking a healthy gulp of water through the straw. What happens? Water drains from between the ice cubes and leaves the system through the straw. The volume of water that left the system divided by the drop in water level and the cross-sectional area of the glass is the unconfined storativity (Fig. 3.7). In other words, the unconfined storativity is the drainable fraction of a unit volume of the aquifer.

Note that the unconfined storativity is less than the porosity because water is left behind due to capillary forces (and the smaller the grain size for the geologic media that makes up the aquifer, the more water is left behind). For sands and gravels, about 15 percent of the volume of an aquifer is drainable, and for sandstone it's about 5 percent. For fractured and karst systems, it may be 1 percent or less.

Unconfined storativity is important when we consider water-level declines due to groundwater production. If we remove a set volume of water from a fractured aquifer, the water-level decline will be about 5

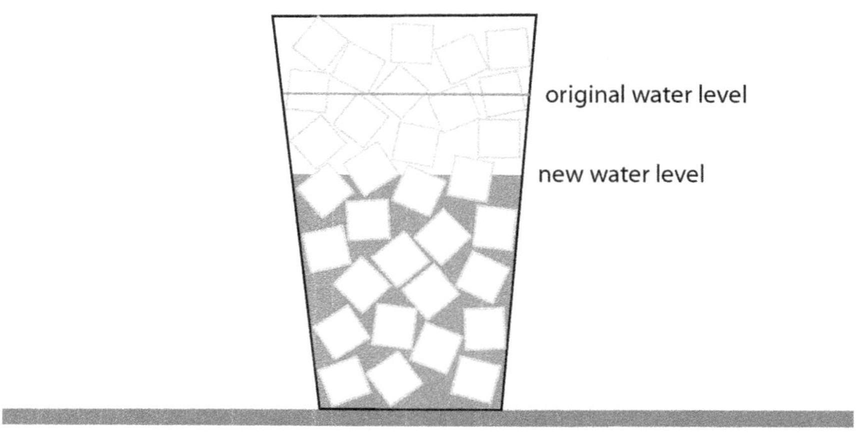

Fig 3.7 Change in unconfined storage in a glass of water with ice

times more if that same amount of water was pulled from a sandstone and about 15 times more that of a gravel. Similarly, the same amount of water produced from a sandstone will produce a water-level decline about three times as large as a gravel or sand. Geology (and unconfined storativity) matters.

In a confined aquifer, the aquifer generally remains fully saturated after water is produced from it. So where does the water come from? It comes from the compressibility of the geologic media and the water in it. I don't know if you hug rocks or not, but if you do, you know that they are not terribly huggable. In other words, they are not terribly compressible. Furthermore, water is generally assumed to be an incompressible fluid. But geologic media and water are, indeed, slightly compressible but enough to mean something in aquifers. The compressibility of water is 4.4×10^{-10} per pascal where pascal is a measure of pressure named after Blaise Pascal (1623 to 1662 AD), who dabbled in fluid dynamics and many other topics. That isn't much, but it's something. The compressibility of sand is about 10^{-7} to 10^{-9} per pascal while the compressibility of fractured rock is about 10^{-9} to 10^{-10} per pascal. Clay is more compressible than sand at 10^{-6} to 10^{-8} per pascal. As you can see, water is generally less compressible than geologic media, but, depending on the geologic material, can still comprise a large part of overall aquifer compressibility.

We mentioned earlier that water in a confined aquifer is under pressure. That pressure is, in part, helping to hold up the rocks and sediments above it. Rocks may not be huggable, but they sure are heavy. Sitting at sea level, you feel about 100 kilopascals of pressure. At 30 meters below the top of a lake, you feel about 1200 kilopascals of pressure. Under 30 meters of saturated sandstone, you would feel 13,000 kilopascals of pressure. So, in a confined aquifer, when the local water pressure is decreased, all that weight bearing down from above compresses the aquifer a small amount more, which provides a small volume of water. In addition, with the decline in pressure, water expands a slight amount, also providing a small volume of water, all while the aquifer remains fully saturated with water.

To use a different fluid and situation as an analogy, think about the air in a tire on an automobile. Where is the pressure coming from in the tire? In part, it is from the air in the tire, but it's also in part due to the weight of the car bearing down on the air in the tire. Now let's release half of the pressure out of the tire (this is a good experiment to conduct on your neighbor's car instead of your own). What's the result? One, the weight of the car has compressed the tire somewhat (it looks flatter), but the tire is still holding (and is full of) air. Sure, the tire is not as full of air as it was before we let air out of the tire, but it is still full of air. The same is true of water in confined aquifers.

The release of water from a confined aquifer is generally small compared to the release of water from the drainable storage in an unconfined aquifer. The storativity in confined aquifers ranges from 0.005 to 0.00005. These values are directly comparable to storativity values in an unconfined aquifer (0.01 to 0.15). This means that a much greater volume of a confined aquifer is affected by a set amount of groundwater production than in an unconfined aquifer. Not as much of an unconfined aquifer has to be drained as a confined aquifer has to be depressurized and compressed to provide the same volume of water. As we will see, this is a critical point for understanding how aquifers respond differently to groundwater production depending on their hydrologic setting, namely whether they are confined or unconfined.

Water-Level Declines from Production

So far, we have used a glass of water as an analogy for an aquifer (as well as your neighbor's car), but, indeed, aquifers are far more complicated (and immense) than a glass of water. While the concepts of porous media, water-level declines, and storativity are the same, real aquifers respond differently than a glass of water.

When water is produced from an aquifer—pumped from an unconfined aquifer and pumped or allowed to flow in a confined aquifer (if pressure pushes the water above land surface), the water level in the well, and thus the aquifer, declines in response to that production from the ***static water level***, the water level in the aquifer before production. This water-level decline associated with production is called ***drawdown***. Water flowing to a producing well is simply water flowing from higher

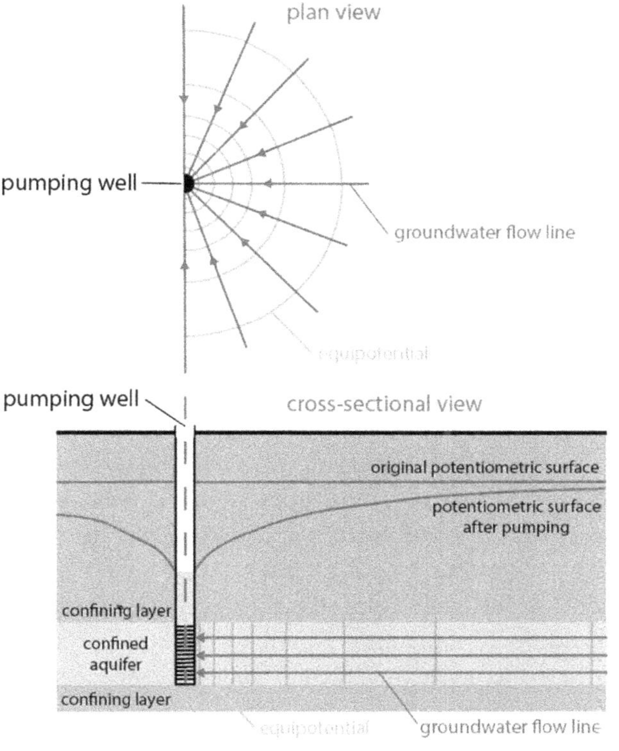

Fig. 3.8 Water-level declines (drawdown) around a producing well (from Freeze and Cherry 1979, p. 316)

potential energy (a higher water-level elevation) to lower potential energy (a lower water-level elevation). Because a well is a point in what amounts to a two-dimensional layer and, with production, we have lowered the water level at the well, groundwater flows toward the well from all lateral directions[5] (Fig. 3.8).

Unlike our glass of water where the water level decreased uniformly across the cross-sectional area of the glass, water levels in an aquifer almost always

[5] There can be complexities here with partially penetrating wells, boundary conditions, and heterogeneity; I am discussing the general, most common occurrence.

decline from a background water-level surface in the aquifer.[6] This is due in large part to the high permeability of ice in a glass compared to that in real-world aquifers. Measured laterally out from the well, the highest water-level decline (lowest water-level elevation, at least locally) occurs at the well with lower water-level declines as we move away from the well until we reach background, preproduction water levels (Fig. 3.8). The decline of water levels around a well is also not linear: the magnitude of the declines is larger closer to the well than farther away. When seen in three-dimensions, the water-level declines around a producing well look like a cone, so it is known as the ***cone of depression*** (depression referring to the decline in water levels, not the unbearable sadness of producing unsustainably from an aquifer).

Production from confined and unconfined aquifers results in water-level declines around the producing wells; however, as noted earlier, water levels in confined and unconfined aquifers respond differently to production. For example, if we have a sandstone with the same permeability and thickness and produce water from it at the same rate for the same period of time, the water-level decline at the production well of a confined aquifer would be 13.4 meters while production from the same aquifer if it was confined would be 22.2 meters[7] (Fig. 3.9). We can also calculate the lateral size of the cone of depression for our sandstone aquifer given this production. For the unconfined aquifer, the cone of depression extends 50.6 meters from the production well. For the confined aquifer, the cone of depression extends 878 meters from the production well.

The reason the cone of depression in the confined aquifer is so much deeper and broader than the cone of depression for unconfined aquifer is the difference in storativity. The cone of depression in the unconfined aquifer is so much smaller because the water from storage comes from physically draining the aquifer, resulting in a large storativity. The cone of depression in the confined aquifer is so much larger because the water sources from compressibility of the aquifer and water, which results is a

---

[6] Karst aquifers can be an exception here. For example, it is not uncommon for a well in the Edwards Aquifer of Texas to be pumped with no measurable decline in water level in the well (Hovorka and others 1998).

[7] I am using the Theis (1935) equation to calculate these water-level declines. This equation mathematically describes water-level declines over time around a produced well in a confined aquifer. The equation can also be used to estimate early-phase water-level declines around an unconfined well as long as the drawdown in the well is small compared to the thickness of the aquifer.

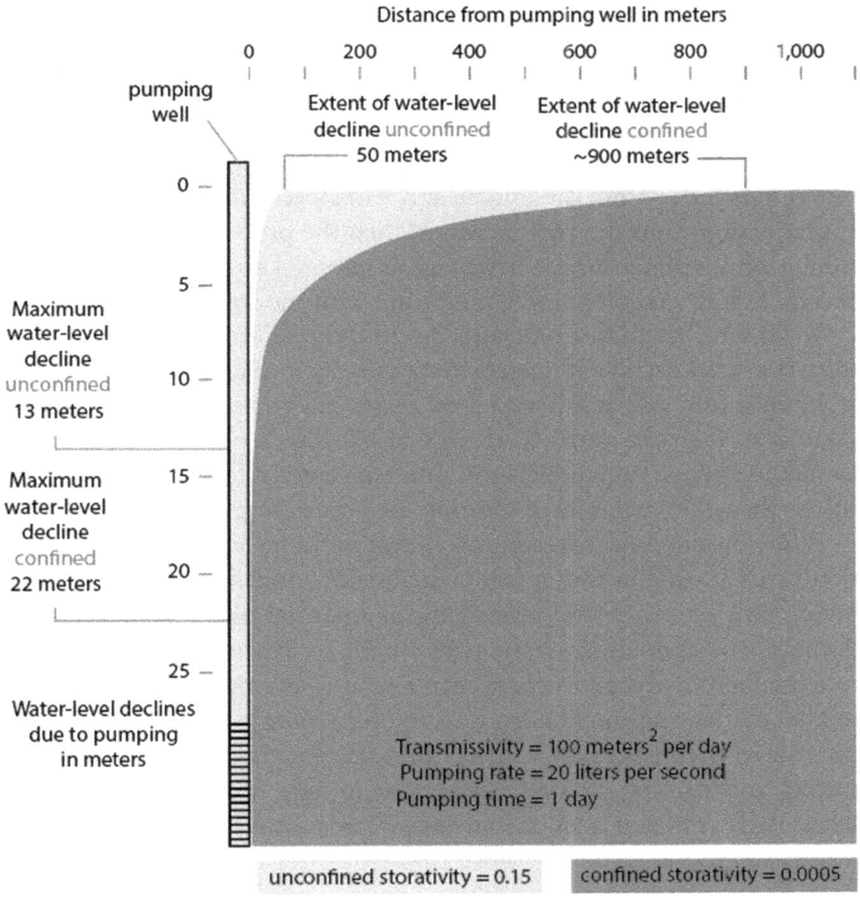

Fig. 3.9 Water-level declines (drawdown) around a well with the same properties and pumping scenario but under confined and unconfined conditions

much smaller storativity. Just like zombies wander the countryside in search of brains (brains!), the cone of depression radiates outward from a producing well in search of water in storage.

When producing from an aquifer, the water initially sources from a change in storage as reflected in the decline in water levels. As the cone of depression gets larger, the rate of water-level decline at any single point decreases with time although the overall change in storage with time remains the same. At some point, the cone of depression captures either discharge, say to a spring or river, or recharge coming into the aquifer. In a sense, recharge and discharge are one and the same as they represent natural flow through the aquifer intercepted by production. If there is enough catchable discharge and recharge, then the production will come into *equilibrium* within the aquifer, and water levels will no longer change. If production from the well increases, then the process to reach equilibrium begins again: a change in storage as reflected by water-level declines and then slowly coming into equilibrium with the aquifer, if possible.

Water-level declines from wells are linearly additive. For example, if the water-level decline 30 meters from a single producing well is 2 meters at equilibrium, and another well 60 meters away starts pumping at the same rate, then the water-level decline halfway between the two wells would be 4 meters at equilibrium. This concept is referred to as superposition, the additive nature of multiple wells pumping simultaneously.

When production comes into equilibrium in an aquifer, that production is sustainable in that it can be maintained indefinitely at that volumetric rate with that level of recharge and natural discharge. If recharge declines, the cone of depression may need to expand to capture more recharge and discharge to come into a new equilibrium. It is important to note that any production from an aquifer results in some magnitude of water-level decline and, eventually, some decrease in natural discharge to springs and streams. It is also important to note that in some aquifers, particularly confined ones, it may take decades to centuries (or more) to reach equilibrium.

Impacts of Water-Level Declines

Water-level declines in and of themselves may seem innocuous, but they can impact well productivity, baseflow to streams and springs, water quality, and even the aquifer itself. Declining water levels can decrease, or even dry up, flows in flowing wells, thus impacting those water sources and requiring water users to install and use pumps. If water levels are below the land surface, then lower water levels increase pumping costs since the water must be lifted to the land surface.

In confined aquifers, large-capacity wells, generally for municipalities, may have different borehole diameters with depth. The initial borehole may be a certain diameter to access the aquifer. The borehole may then be reamed (redrilled) for part of its depth to a larger diameter to accommodate a pump. For example, if a confined aquifer is between 400 and 500 meters in depth and the depth to water is 100 meters, an initial borehole of 0.3 meters in diameter may be drilled to 500 meters in depth. Then the borehole may be reamed to 0.6 meters in diameter down to 300 meters in depth. Partial reaming is done to save money. If water levels decline too much relative to the pump elevation, the user may have to rework the well (ream it deeper) or—more likely—drill a new well to achieve the same pumping rates.

If water levels continue to decline, they may affect the ability to produce groundwater at previous rates. For unconfined aquifers, the saturated thickness decreases as water levels decrease. At some point, the aquifer may no longer be able to accommodate the drawdown from previous pumping rates, forcing the user to reduce their production rate. The same issue occurs in a confined aquifer, especially if previous production rates relied upon large artesian water-level declines to achieve pumping rates. In cases where the aquifer permeability is high, an aquifer may have to change from confined to unconfined before groundwater producers experience a decrease in production rate.

Eventually, water-level declines due to production may interfere with aquifer recharge and discharge. For recharge, from the perspective of the aquifer, water-level declines may cause an increase in recharge if water-level declines change a gaining stream into a losing stream. While this might be good from the aquifer's perspective, it might adversely affect surface-water users (including the environment) from decreasing or even drying up the stream. For discharge, groundwater production—to come into a new equilibrium—may intercept groundwater flow that would have contributed to springs and streams if the pumping wasn't there. This might be good from the groundwater perspective but, similar to changes to recharge processes, it might adversely affect surface-water users (including the environment) from decreasing or even drying up the streams and springs.

If groundwater production is large enough, it may not be possible to capture enough recharge and discharge for the aquifer to come into equilibrium. In this case, production may need to be met with water from storage meaning that water levels may continue to decline over time since they cannot come into equilibrium. This is referred to as *aquifer mining*, *groundwater mining*, or simply *mining*.

Water-level declines may induce changes in groundwater quality. For aquifers bordering the sea, water-level declines may induce seawater intrusion, spoiling freshwater parts of the aquifer. For islands, there is often a lens of freshwater floating on top of saline water such that groundwater production not only creates a cone of depression in the aquifer but also creates a cone of impression (upwelling) from underlying saline water. More locally, water-level declines might induce poorer quality water to move into the freshwater parts from bordering confined layers or interbedded layers of clays and shales. In unconfined aquifers, a lowering of the water table may pull human-contaminated groundwater into deeper-screened wells.

In some cases, particularly in aquifers made of karst limestone or loose sediments that include clay, water-levels declines may cause the aquifer to partially collapse on itself. In karst aquifers, water-level declines may cause sinkholes to form (which may be amplified by drought and development). In aquifers made of unconsolidated sediments, water-level declines may lead to broad land subsidence, increasing flooding, the risk of seawater intrusion, changing surface-water flows, and damaging property.

The Bottom of the Borehole

In the end, the question of groundwater sustainability is answered not only through balancing groundwater production with flows in the aquifer but also with the impacts of water-level declines on surface water, the aquifers, and all its users, including the environment. Early in human history, we used aquifers sustainably because we either relied on what they gave us through springs or baseflow to streams or used wells that couldn't produce much water. However, once technology allowed high production rates from wells, humans began testing and exceeding the sustainable limits of aquifer. Even with the sustainable production of groundwater, groundwater production may impact water levels in the aquifer and, ultimately, discharge to springs, streams, and other surface-waters.

Questions

Do you have any springs in your area? How much do they flow?
Have you ever seen a well? Describe it.
Are water levels in your area stable or declining?
What have been the effects of water-level declines in your area?

References

Agricola, G., 1556, *De re metallica*: 1st edition, Basil, Hieronymus Froben & Nicolaus Episcopius, 594 p.

American Journal of Science, 1853, v. XV.

Baker, T.L., 1985, *A Field Guide to American Windmills*: University of Oklahoma Press, 516 p.

Bell, T.J., 1881, History of the water supply of the world: Cincinnati, Ohio, Peter G. Thomson, Publisher, 134 p.

Bullard, L., 2004, Healing waters – Missouri's historic mineral springs and spas: University of Missouri Press, 243 p.

Calderone, L., 2018, Using windmills to deliver water: AgriTech Tomorrow, posted March 13, 2018, https://www.agritechtomorrow.com/article/2018/03/using-windmills-to-deliver-water/10595/.

Chondros, T.G., 2010, Archimedes influence in science and engineering *in* Paipetis, S.A., and Ceccarelli, M., editors, *The Genius of Archimedes – 23 Centuries of Influence on Mathematics, Science and Engineering, History of Mechanism and Machine Science*, Springer Science+Business Media, p. 411–425, https://doi.org/10.1007/978-90-481-9091-1_31.

Clark, G., 1944, Antiquity: v 18, n 69, p 1–15.

Combes, M., 1838, Surles Roues de Reaction: Académie des Sciences, v 7.

Combes, M., 1840, Recherches expérimentales sur les roues à réaction: Académie des Sciences, v 10, p 362.
Conner, C.D., 2005, A people's history of science: Nation Books, 554 p.
Cox, W.G., 1895, Artesian wells as a means of water supply: Sapsford & Co., 146 p.
de Feo, G., Angelakis, A.N., Antoniou, G.P., El-Gohary, F., Haut, B., Passchier, C.W., and Zheng, X.Y., 2013, Historical and technical notes on aqueducts from prehistoric to medieval times: Water, v 5, p 1996–2025; https://doi.org/10.3390/w5041996.
Degousée, J., 1847, Guide du sondeur, ou Traité théorique et pratique des sondages: Langlois et Leclercq, Paris, Volume 1, 504 p.
de Saint-Exupery, A., 1943, *The Little Prince*: Reynal and Hitchcock, New York.
Deutsches Museum, 2021, Roman double-piston pump: accessed April 11, 2021, https://www.deutsches-museum.de/en/collections/machines/power-engines/water-power/piston-pumps/double-piston-pump/.
Fattahi, M., 2015, OSL dating of the Miam Qanat (KARIZ) system in NE Iran: Journal of Archaeological Science, v 59, p. 54–63.
Freeze, R.A., and Cherry, J.A., 1979, Groundwater: Prentice-Hall, Inc., 604 p.
Garfinkel, Y., Vered, A., and Bar-Yosef, O., 2006, The domestication of water—The Neolithic well at Sha'ar Hagolan, Jordan Valley, Israel: Antiquity, v 80, p 686–96.
Goes, B.J.M., Parajuli, U.N., Haq, M., and Wardlaw, R.B., 2017, Karez (qanat) irrigation in the Helmand River Basin, Afghanistan – A vanishing indigenous legacy: Hydrogeology Journal, v 25, p 269–286.
Green, D.E., 1973, *Land of the Underground Rain—Irrigation on the Texas High Plains, 1910–1970*: University of Texas Press, 295 p.
Hallenburg, J.K., 1997, Introduction to geophysical formation evaluation: CRC Press, 181 p.
Harris, L.E., 1953, Some factors in the early development of the centrifugal pump 1689 to 1851: Transactions of the Newcomen Society, v 28, n 1, p 187–202, https://doi.org/10.1179/tns.1951.014.
Hills, R.L., 1994, *Power from Wind – A History of Windmill Technology*: Cambridge University Press, 324 p.
Hovorka, S.D., Mace, R.E., and Collins, E.W., 1998, Permeability structure of the Edwards Aquifer, South Texas – Implications for aquifer management: Bureau of Economic Geology, The University of Texas at Austin, Report of Investigations No. 250, 55 p.

Jansen, M., 1989, Water Supply and Sewage Disposal at Mohenjo-Daro: World Archaeology, v 21, n 2, p 177–192.

Jung, L., and Fang, L.-C.T., 1948, An Account of the Salt Industry at Tzu-liu-ching *chi*: Isis, v 39, n 4, p 228–234.

Lane, K., 2017, Puquios and aqueducts in the Central Andes of South America: Boca Raton, Florida: CRC Press. p. 465–467, *in* Underground Aqueducts Handbook.

le Demour, K., 1732, Machines approvées: L'Academie Royale des Sciences de Paris, v 6, n 363, p 9.

Litton, A., 1860, Belcher & Brother's artesian well: Transactions of the Academy of Science of Saint Louis, v. 1, p. 80–86.

Lynch, P.N., Shepard, C.U., and Geddings, J.F.M., 1884, Artesian well – Charleston's water supply: The Sanitarian, Medico-Legal Society of New York, v. 12, p. 110–127.

Mackworth, H., 1854, Improvements in boring for minerals, wells, and shafts: Transactions of the North of England Institute of Mining and Mechanical Engineers, v. 2.

Mays, L.W., 2010, Water technology in ancient Egypt *in* Mays, L.W., editor, *Ancient Water Technologies*: Springer Science and Business Media. B.V.: Dordrecht, The Netherlands, Chapter 3, p 53–66.

Meinzer, O.E., 1934, History and development of ground-water hydrology: Journal for the Washington Academy of Sciences, v 24, n 6, p 6–32.

Meunier, M., 1978, Les puits artésiens de l'Avallonnais: Académie du Morvan, v. 5, n. 6, p. 3–14.

Merriam-Webster, 2021, History and Etymology for well: https://www.merriam-webster.com/dictionary/well.

Papin, D., 1689, Rotatilis suctor et Pressor Hassiacus, in serenisfima aula caffellana demonstratus & detectus: Actis Eruditorum, June, p 317–322.

Peltenburg, E.J., Croft, P., Jackson, A., McCartney, C., and Murray, M.A., 2001, Well-Established Colonists: Mylouthkia 1 and the Cypro-Pre-Pottery Neolithic B *in* Swiny, S., (editor), *The Earliest Prehistory of Cyprus: From Colonization to Exploitation*, Swiny, American Schools of Oriental Research Archaeological Reports 5; Cyprus American Archaeological Research Institute Monograph Series 2. Boston: American Schools of Oriental Research, p 61–93.

Rybníčeka, M., Chlupc, T., Kalábekd, M., Kalábkováe, P., Kočárf, P., Kynclg, T., Muiggh, B., Tegelh, W., Vostrovskáe, I., and Kolářa, T., 2018, New dendroarchaeological evidence of water well constructions reveals advanced Early Neolithic craftsman skills: Dendrochronologia, v 50, p 98–104. https://doi.org/10.1016/j.dendro.2018.06.003

Shepherd, D.G., 1990, Historical development of the windmill: NASA Contractor Report 4337, 43 p.

Shufeldt, G.A., 1867, History of the Chicago Artesian Well – A demonstration of the truth of the of the spiritual philosophy with an essay on the origin and uses of petroleum: Religio-Philosophical Publication Association, 58 p.

Steinel, A.T., 1926, *History of Agriculture in Colorado – A Chronological Record of Progress in the Development of General Farming, Livestock Production and Agricultural Education and Investigation, on the Western Border of the Great Plains and in the Mountains of Colorado, 1858 to 1926*: Colorado Agricultural College, Fort Collins, 659 p.

Streit, K., and Garfinkel, Y., 2015, A specialized ceramic assemblage for water pulling – The Middle Chalcolithic well of Tel Tsaf, Israel: Bulletin of the American Schools of Oriental Research, v 374, p 61–73.

Swindell, J.G., and Burnell, G.R., 1883, Rudimentary treatise on wells and well-sinking: Crosby Lockwood and Co., London, 166 p.

Taghavi-Jeloudar, M., Han, M., Davoudi, M., Kim, M., 2013, Review of ancient wisdom of qanat, and suggestions for future water management: Journal of Environmental Engineering and Science, v 18, n 2, p 57–63.

Theis, C.V., 1935, The relation between the lowering of the piezometric surface and the rate and duration of discharge of a well using groundwater storage: Transactions of the American Geophysical Union, v 16, p 519–524.

Tolman, C.F., 1937, *Ground Water*: McGraw-Hill Book Co., Inc., New York, N. Y., 1937, 593 p.

Van Esch, R.P.M., 1997, Simulation of three-dimensional unsteady flow in hydraulic pumps: Ph.D. dissertation, University of Twente, Enschede, Nederlands, 147 p.

Villiers, Marq de, 2003, Water, Random House, 464 p.

Vitruvius (Marcus Vitruvius Pollio), n.d. (published ~30-15 BC), *De Architectura*, Book X.

Voudouris, K., Valipour, M., Kaiafa, A., Zheng, X.Y., Kumar, R., Zanier, K., Kolokytha, E., and Angelakis, A., 2019, Evolution of water wells focusing on Balkan and Asian civilizations: Water Supply, v 19, n 2, p 347–364.

Weiner, J., 1994, Well on my back – An update on the Bandkeramik wooden well of Erkelenz-Kückhoven: NewsWARP, v 16, p 5–17.

Wells, D.A., Cross, C.R., Bliss, G., Trowbridge, J., Kneeland, S., and Nichols, R., 1854, Annual of scientific discovery – Or, year-book of facts in science and art: Gould and Lincoln.

Whitney, J.D., 1894, The United States—Facts and figures illustrating the physical geography of the country, and its material resources: Supplement I. Population: Immigration: Irrigation, Little, Brown, and Company, Boston, 324 p.

Wilkinson, T.J., Boucharlat, R., Ertsen, M.W., Gillmore, G., Kennet, D., Magee, P., Rezakhani, K., and Schacht, T., 2012, From human niche construction to imperial power – Long-term trends in ancient Iranian water systems: Water History, v 4, p 155–176.

Yannopoulos, S.I., Lyberatos, G., Theodossiou, N., Li, W., Valipour, M., Tamburrino, A., and Angelakis, A.N., 2015, Evolution of water lifting devices (pumps) over the centuries worldwide: Water, v 7, p 5031–5060; https://doi.org/10.3390/w7095031.

4

From Safe Yield to Where?

"the growing popularity of ground-water supply for irrigation and the heavy drafts made possible by improved pumping machinery and cheap power are giving rise to conditions of dangerous overdraft on many of the so-called inexhaustible underground water supplies."
Charles H. Lee (*1915*)

Before humans started producing water, aquifers were in a ***dynamic equilibrium***. In other words, water flowing into and out of aquifers were in a state of balance over time. Water levels and springflows might rise during wet periods and fall during droughts based on the amount of recharge (which ties back to precipitation), but, on average, the system was in equilibrium. An aquifer would also be in a dynamic equilibrium during periods of drought and climate change. For example, a drought or drying climate would decrease recharge which would, in turn, decrease discharge a corresponding amount.

For thousands of years, human impacts on aquifer flows and volumes were minimal. Using springflows had no impact on the aquifer unless springs were enhanced (such as through clearing obstructions or lowering the elevation of the discharge point to increase flow). Although early artificial removal of water from an aquifer such as through spring enhancement, hand-dug wells, qanats, and windmills had impacts on the flow

R. E. Mace, *Groundwater Sustainability*, Palgrave Studies in Environmental Sustainability, https://doi.org/10.1007/978-3-031-13516-3_4

system, ultimately decreasing discharge somewhere in the aquifer, these interventions were generally miniscule compared to the overall volume of flow through an aquifer. This early use of aquifers was ***sustainable***, which I define here as being able to produce water from an aquifer at a defined volume in perpetuity under relevant recharge conditions. I add the qualifier "under relevant recharge conditions" to account for climate change, land-use change, artificial recharge, or anything else that may impact the entry of water into the aquifer.

Tapping into and bleeding off the pressure of artesian aquifers that produced flowing wells was probably the first time humans were able to induce aquifer-wide water-level declines that affected the ability for people to indefinitely use their aquifers. As shown at the beginning of Chap. 1, Texans disregarded the science and signs of trouble until flowing artesian wells across the aquifer were reduced to a trickle, an experience shared with well-owners in California's Owens and Central valleys and elsewhere across the world. This is known in economics as the ***tragedy of the commons*** where the system compels the individual to increase their use of a shared but limited resource resulting in overuse (Hardin 1968).

Once the artesian pressures that induced flow at land surface were depleted, those aquifers had a reprieve until the invention of the downhole centrifugal pump, especially once the diameter of this pump became small enough for farmers to afford to drill wells and install them. Able to produce 60 liters or more per second, hundreds to thousands of these wells were able to produce a volume of water rivaling and even exceeding natural flows through aquifers. Thus began a more than one-hundred-year discussion that continues today among various specialists on how to best govern and manage groundwater.

The next four chapters will explore this topic in detail. While I could have just jumped to the conclusion and presented my preferred definitions and been done with it, I instead decided to present the full history of the topic. Understanding the history of what happened helps to explain why things are the way they are today and perhaps opens doors to new solutions. This chapter focuses on the introduction of the term "safe yield" and the subsequent debates—sometimes fiery—on the term through the 1970s, up to when global discussions on the general topic of sustainable development started. The following chapter focuses on the tension between sustainable and nonsustainable production. The chapter after that returns to the safe yield discussion but under the broader umbrella of sustainable development. I follow that chapter with a discussion of legal doctrines and governance associated with sustainable management of aquifers.

In this chapter, we introduce the perspective of engineers and hydrogeologists on groundwater management. The earliest engineering and hydrogeologic thoughts were to manage sustainably, although the focus was on maximizing production from aquifers for the sole benefit of human use. This discussion focused on the ***safe yield*** of an aquifer (Curtis et al. 1979), the volume of water that can be perennially produced from an aquifer without producing undesirable results (modified from Todd [1959]).

Safe Yield

Since the management of rivers preceded the management of aquifers, it is not surprising that groundwater management initially borrowed concepts and terms from surface-water management. One such term is safe yield. Surface-water engineers and managers use a term called yield to manage surface-water reservoirs. *Yield* is a term that quantifies the capacity of a water resource to serve as a long-term water supply (Leib and Stiles 1998). *Firm yield* for surface water is the maximum amount of water that can be annually produced from a reservoir during a repeat of the drought of record. Firm yield is generally presented as a volume over a year, such as megaliters per year.

A firm yield is not a guarantee that the water will be there. Changes in the watershed, sedimentation in the reservoir, a drought more severe than the drought of record, and a limited record would affect the firmness of the firm yield. Furthermore, during an exact repeat of the drought of record, the reservoir goes dry on the last day of the drought (or, more accurately, its conservation storage becomes fully depleted), something that might be disturbing to water managers and users. Therefore, surface-water engineers and managers may use a *safe yield* for planning, regulatory, and management purposes. The safe yield anticipates a drought worse than a drought of record, usually defined as providing the yield for a certain amount of time longer than the drought of record. In essence, safe yield builds in a safety factor to firm yield, although it, too, may be subject to changed conditions.

Building on his earlier work where he conducted a detailed water-budget analysis of a closed-basin, alluvial aquifer California's Owen Valley where he discussed groundwater yield as being equal to recharge (Lee 1912), Lee (1915) was the first to use the term safe yield for a groundwater system. Lee (1915) provided two inter-related definitions of safe yield: (1) "the net annual water supply which may be developed by pumping and Artesian flow without persistent lowering of the ground-water plane" and (2) "the limit to the quantity of water which can be withdrawn regularly and permanently without dangerous depletion of the storage reserve." The groundwater plane he refers to is the potentiometric, or water-level, surface. The dangerous depletion he refers to in the second definition is the persistent lowering of water levels resulting in chronic depletion of groundwater in storage. Avoiding this persistent lowering of water levels meant that Lee's definition of safe yield would result in sustainable use of the aquifer.

The safe yield of a groundwater reservoir, as defined by Lee and hereafter by many others, is fundamentally different than the safe yield of a surface water reservoir. In a surface-water reservoir, there are in-flows to the reservoir, possibly natural out-flows from the reservoir, a volume of storage, and an annual yield. That annual yield is based on the in-flows and the storage with the totality of the usable storage consumed during a defined time period under drought conditions. In a groundwater system, there are in-flows to the reservoir, natural out-flows from the reservoir, and an annual yield. Although there are changes to the amount of water in storage in a groundwater reservoir as the aquifer arrives at a new dynamic equilibrium with lower water levels and decreased discharge, total groundwater storage is, in almost all cases, irrelevant.

Lee (1915) presciently focused his definition of safe yield on aquifer conditions, specifically stable groundwater levels, despite writing about an aquifer that perhaps comes the closest to behaving like a bathtub from the water budget perspective. And he recognized that safe yield was less than the recharge rate.

Lee (1915) developed a water budget for the basin and correctly noted that increases in production resulted in decreases in aquifer outflows, what he referred to as "natural groundwater loss," and the drying up of springs and *ciénegas*, the latter of which are wetlands unique to the American Southwest. Lee (1915) also discussed how to increase the safe yield of aquifers through (1) "the reduction of residual losses to the lowest possible quantity," that is, drying up springs and ciénegas; (2) conserving groundwater (Owens Valley also struggled with people leaving their flowing wells flowing for no reason); and (3) increasing recharge through flood waters. Lee (1915) noted that the theoretical limit for safe yield was the recharge rate but that in practice it was less than recharge, including economic limits of production. Lee (1915) also noted that groundwater in storage can be used during droughts and presumably replaced during wetter times.

Lee (1915) restricted his conclusions to closed-basin systems where recharge occurred on basin boundaries where aquifer sediments lapped up against mountain ranges and natural discharge occurred to springs and wetlands in the basin. This is something of a special case where the use of groundwater would not impact surface-water rights, although it would impact surface-water discharge and, thus, have environmental impacts, something that was not a concern at that time.

Lee's 71-page paper was greeted with 33 pages of discussion from six authors, all of whom were complimentary of his paper, correctly identifying it as pioneering work. Smith (1915) proposed that Lee's ideas and concepts could be applied to a broader range of geologic and climatic conditions, including open basins, and noted that "groundwater loss" from springs could be prevented through pumping (i.e., drying up the springs) and that losses from evapotranspiration could be prevented by lowering water levels to below root zones. Smith recognized that "outlet loss"—the flow of groundwater out of the basin—was, in the bigger picture, not a loss "to the State" because one basin's loss was another basin's gain. Smith also noted that, depending on recharge, the safe yield could vary from year to year and that practitioners probably needed to seek average conditions, perhaps through estimations over several years.

O.E. Meinzer (1915), one of the godfathers of hydrogeology, noted that "a proper distribution of wells is necessary in order to reduce" losses from springs, ciénegas, and evapotranspiration and that losses might be unavoidable due to the quality of land where the losses occur (in other words, there might not be a local use of pumped groundwater at natural discharge points).

Allen (1915) noted the uncertainty in estimating safe yield due to the uncertainty in measuring various parts of the groundwater budget. Meinzer (1915) made a similar comment in expressing disappointment that Lee did not discuss the probable accuracy of his results. Owen (1915) simply wanted to warn the world away from using groundwater due to its unreliability and need for much land for many wells.

As a sign of the times, the author and commentators focused on maximizing the yield from the aquifer to the detriment of surface-water/groundwater interactions and the environment. Given that the aquifer was a closed basin and (presumably) humans were not using the discharge to springs and ciénegas to support economic activity, Lee (1915) was not necessarily proposing that the maximization of using groundwater resources should supersede those of surface water, but that is what carried forward in subsequent decades of discussions on groundwater management. Regardless of the details of his approach—that was a different time with respect to environmental impact—the safe yield he proposed was sustainable.

Firming Up Safe Yield

O.E. Meinzer cemented the use of safe yield in groundwater science, engineering, and management by including it in his paper "Quantitative methods of estimating ground-water supplies" (Meinzer 1920) and two of his influential U.S. Geological Survey reports "Outline of Ground-Water Hydrology—Definitions" (Meinzer 1923) and "Outline of Methods for Estimating Ground-Water Supplies" (Meinzer 1932). But he did change the definition, and not necessarily for the best.

Meinzer's 1920 paper equated safe yield to recharge, possibly starting a misconception that still exists today (see discussion in Chap. 6 on the Water Budget Myth). His 1923 U.S. Geological Survey report defined safe yield as "the rate at which water can be withdrawn from an aquifer for human use without depleting the supply to such an extent that withdrawal at this rate is no longer economically feasible." This is an interesting definition because it suggests that water levels should not be drawn down beyond which it is no longer affordable to produce. This may also be his attempt to develop a definition that didn't use the dramatic word "dangerous" that Lee (1915) used.

By 1932, Meinzer evolved his definition for safe yield to "the practicable rate of withdrawing water from an underground reservoir perennially for human use." Meinzer (1932) noted that safe yield could be less than or greater than recharge, with the former due to the inability to intercept (salvage is the word he used) all of the natural discharge and the latter due to increased recharge as a consequence of production or through artificial means. Meinzer (1932) also noted, based on work by Wolff (1911) and Smith (1924), that groundwater production could induce and store more recharge by lowering the water table. Finally, due in part to some aquifers already showing systemic water-level declines year after year, Meinzer (1932) concluded that "the most urgent problems in ground-water hydrology at the present time are those relating to the rate at which rock formations will supply water to wells in specified areas—not during a day, a month, or a year, but perennially."

Tolman (1937) noted that if the distance between the recharge zone and groundwater production was large, as seen in some of the artesian basins, the safe yield of the system related more to the "capacity of the formation as a conduit rather than to its capacity as a reservoir." A somewhat extreme example of this is the Trinity Aquifer in Texas north of the Colorado River. Discussed in the introduction to Chap. 1, the Trinity Aquifer consists of three sandstone beds that rise to the land surface west of Dallas, Fort Worth, and Waco, and slowly dive into the subsurface toward the Gulf of Mexico.

Mace and others (1994) showed that, although artesian water levels had declined as much as 335 meters in the aquifer in the Dallas-Fort Worth area—increasing the down-dip hydraulic gradient threefold—there had not been a measurable decline in water levels or groundwater discharge to streams and springs in the outcrop. That meant that the amount of groundwater moving deeper into the aquifer toward pumping centers was much lower than the recharge to the aquifer, the vast majority of which naturally discharged to streams and springs in the outcrop. Many years later, a groundwater model showed that of the 2.2 million megaliters per year that recharged the Trinity and Woodbine aquifers in predevelopment times, only 0.8 percent of it flowed into the confined part of the aquifer (Kelley and others 2014). Just as Tolman (1937) observed for a general artesian aquifer, safe yield of the Trinity Aquifer relates more to the ability of the Trinity Aquifer to transmit water than to its recharge rate.

A classic paper by Theis (1940) expanded on Meinzer's (1932) description of what happens when water is produced from wells and summarized prior thoughts about safe yield. Theis (1940) noted that, prior to groundwater development through wells, aquifers were in a dynamic equilibrium where the average rate of discharge from an aquifer was equal to the average rate of input (recharge and cross-formational flow). Theis (1940) further stated that discharge from wells had to be balanced by either an increase in recharge, decrease in natural discharge, a change in storage, or a combination of them. He further noted that parts of the aquifer in the recharge zone may be full and thus reject potential recharge. Therefore, recharge can be increased by lowering water levels and converting groundwater discharge areas to groundwater recharge areas. If an aquifer does not reject recharge, then there is no potential for increasing recharge through increased pumping.

Theis (1940) stated that it is sometimes assumed that recharge is a measure of the possible yield to wells but notes that wells are not generally able to stop all natural discharge. He also identified a change in storage due to a change in water level as a source of water to a well. Finally, he stated that the "ideal development of any aquifer from the standpoint of the maximum utilization" is to (1) place wells as close to rejected recharge areas and discharge areas as possible, (2) place wells uniformly in areas between the rejected recharge zone and the discharge zone, and (3) keep pumping rates below the sum of the natural recharge plus the rejected recharge.

Stuart (1945) defined safe yield as "the maximum rate at which water may be withdrawn without impairing the quantity or quality of the supply," thus introducing water quality as an element to consider when assessing safe yield.

Conkling (1946) defined safe yield as the annual extraction from a groundwater unit which will not (1) exceed the average annual recharge, (2) lower the water table such that the permissible cost of pumping is exceeded, and (3) lower the water table such that water of undesirable quality is allowed to intrude into the aquifer. Conkling identified seawater intrusion and saline groundwater as potential sources of salinity that could be moved into fresher parts of the aquifer due to groundwater production. Conkling also proposed using unused wells for artificial recharge by directing unused surface water into the aquifer, although Sopp (1946) noted the complexities of doing that, including siltation and bacterial growth. Conkling (1946) also noted that consumptive use, rather than pumping volume, should be used in the water budget to account for return flows to unconfined aquifers from irrigation.

Conkling's 30-page article resulted in 51 pages of commentary by 10 readers. Charles (1946) proposed dropping the word "safe" from safe yield since what was safe might be subject to different opinions. Drummond (1946) discussed New Mexican water law where the state's supreme court ruled that the use of groundwater had to be reasonable and avoid injury to others. Baker (1946) discussed the potential legal issues with who owns excess recharged water once it naturally discharges but, oddly, not the corollary of groundwater pumping's impacts to surface water. Forbes (1946) mentioned land subsidence due to groundwater production but from the perspective of a decrease in aquifer storage capacity and not something to consider when establishing a safe yield.

The Unbearable Ambiguity of Safe Yield

Charles (1946) appears to be the first to question the term safe yield, proposing that the profession drop the "safe" part because of the ambiguity of the word. Williams and Lohman (1949) proposed using the term perennial yield, which they defined as "the maximum rate at which water can be salvaged from the natural discharge or added to the recharge, or both."

Thomas (1951), in his book "The Conservation of Groundwater," colorfully described safe yield as "an Alice-in-Wonderland term which means whatever its user chooses" noting that safe yield "will vary widely depending upon the stipulations that govern the determination." He proposed presenting safe yield as a family of curves related to pumping lift, well construction, well locations, and induced recharge from precipitation. He also noted that the process needed to be adaptive and safe yield revised as groundwater development progressed.

McGuinness (1951) used the term "sustained yield" and defined it solely as a function of the properties of the aquifer as constrained by how groundwater was developed, noting that sustained yield could be maximized with proper well placement and pumping. He noted that pumping that exceeds the sustained yield led to mining: the progressive loss of storage as the aquifer is pumped. He pointed to the Ogallala Aquifer beneath the Southern High Plains of Texas as an extreme example of groundwater mining.

Parker (1951) preferred to use perennial yield, restricting it to "the quantity of potable water perennially available in an aquifer without quality impairment." He also notes that safe yield did not involve any "aspect of ethical or justifiable withdrawal," introducing a policy concept, but then did not include these concerns in his definition of perennial yield. Banks (1953) expanded safe yield to protect existing groundwater rights in adjacent basins. Muckel (1955) essentially used the Conkling (1946) definition (without citation) with a few embellishments, including the consideration of dry years when evaluating water-level declines and tying the permissible cost of pumping directly to pumping lifts.

Kazmann (1951, 1956) presented a brutal—and mostly misinformed—attack on safe yield. In his over-the-top but oft-cited paper he published in the Journal of the Irrigation and Drainage Division, Kazmann (1951) concluded that safe yield was not a useful concept because it did not have a single value for a given hydrologic situation.

Kazmann concluded that the Meinzer (1920) definition was too general, although his interpretation of the vagueness was that it did not result in a limit to groundwater production. However, that conclusion is incorrect because studies since 1915—including those by Meinzer—had set limits to production to achieve safe yields. Safe yield is, by name and definition, a defined number (or numbers, if appropriate).

Kazmann presented the following, key question: "For any given water bearing formation, is it theoretically possible to determine a single specific figure which can be termed the 'safe yield' of the aquifer, without reference to the total water resources of the region and adjacent regions?" This question is interesting because it places the management of groundwater resources in a larger context.

One of Kazmann's concerns was that the U.S. Geological Survey, which he defined as a "scientific agency" in a derogatory way (apparently not realizing that Meinzer, not to mention the unmentioned Lee, worked for the Survey as engineers when they did their work), removed the consideration of economics and engineering in a definition used by Williams and Lohman (1949), both geologists with the Survey, in a regional study. Plucking a single definition from a single report by two authors and then lambasting the Survey (and all scientists) over it was disingenuous. The closest thing the Survey had to a sanctioned definition for safe yield was Meinzer's (1932) which, with the word "practicable," included economics and engineering considerations.

Kazmann noted that the transmissivity, the hydraulic gradient, and the induced flow of poorer quality water might prevent the production of the safe yield. These are issues that had already been addressed by others. Meinzer (1915) concluded that a proper distribution of wells would be needed to capture losses from the groundwater system. Tolman (1937) showed that the "capacity of the formation as a conduit" (an alternative way of saying the transmissibility) may be the limiting factor for aquifers where the production was distant from the recharge zone. Stuart (1945) introduced the impairment of water quality to the definition of safe yield, which Conkling (1946) also adopted.

Kazmann went on to attack determining safe yield a priori to pumping, noting that a river hydraulically connected to the aquifer could increase its losses to the aquifer thus increasing the safe yield of the aquifer. Again, Kazmann did not read the literature (or read it closely enough) to realize that increasing recharge was part safe yield from the very beginning (e.g., Lee 1915, Meinzer 1932, and Theis 1940). Kazmann pointed to the Conkling (1946) definition as preferred because it included economics through the "permissible cost" of pumping. Kazmann correctly noted that permissible costs were slippery. For example, the permissible costs for a city to access and pump groundwater are likely much higher than for an irrigator.

Kazmann concluded that "difficulties with the definitions are to be considered a warning that it is necessary to stop redefining 'safe yield' and reexamine the entire situation." Accordingly, Kazmann concluded that safe yield "cannot be reconciled with doctrine of prior appropriation" where surface-water use may interfere with a groundwater right if that molecule would have recharged the aquifer or if a water molecule from a river is intercepted by a well before it discharges to the surface. Kazmann (1956) made the connection between groundwater and surface water in determinations of safe yield, noting that "[i]f all surface runoff in the area overlying an aquifer has been appropriated, a perennial supply cannot be obtained from the ground." Despite these concerns, several states in the United States (Colorado in particular) have managed groundwater under prior appropriation.

Kazmann surmised that "Many aquifers are essentially vast, closed reservoirs which are analogous, on a large scale, to petroleum reservoirs. There is no merit in discussion the 'perennial yield' or 'safe yield' of such an aquifer" and that "unless surface water is available to replace the ground water that has been pumped and is used for that purpose, it is unsound public policy to discuss the 'safe yield' of a water-bearing formation." Here, Kazmann suggested that, unless the recharge is substantial or subject to substantial enhancement, safe yield—or any consideration of sustainable pumping—is wrong. Finally, he concluded that "the concept of 'safe yield' is a fallacious one," recommending that it never be used again and expunged from analysis and law.

Kazmann (1956) provided a shrill—and somewhat misinformed—attack on safe yield, although he makes good points on the vagueness of economic considerations. On the one hand, he presented a preferred definition but then, on the other, presented an argument against safe yield wrapped in an argument against sustainable management of groundwater. Ultimately the management of groundwater resources is a policy decision, regardless of the recharge, economics, or engineering—Kazmann's attempt to use science, economics, and engineering to justify groundwater depletion is fallacious.

In spite of Kazmann's attack and a penchant for people to cite his diatribe to this day, the term safe yield lived on. Todd (1959), in his influential textbook *Ground Water Hydrology* and the most cogent and sophisticated review of safe yield at that time, briefly referenced Kazmann's criticism and dismissed it with the observation that "the chief difficulty arises because [safe yield] is often misinterpreted by persons unfamiliar with ground water hydrology as meaning a fixed underground water supply."

Ouch.

Todd (1959) developed a more refined definition in his textbook: "the amount of water which can be withdrawn from it annually without producing an undesired result." Todd's primary contribution was to generalize the impacts of pumping under the banner of "undesired result." Although his focus was on economic, quality, longevity, and landowners, the phrase "undesired result" was a big enough umbrella to encompass future perceptions of undesired results as desires and policies changed over time. Todd also included a legal component to the assessment of safe yield. Todd (1959) included several methods to assess safe yield in aquifers and noted that he agreed with Conkling (1946) that it is difficult to estimate safe yield in a basin without first producing water from it.

Mann (1961, 1963), while noting that safe yield was a function of assumptions, loosely connected it to policy when he wrote that safe yield "may change with changing cultural conditions on the surface of the basin" although it appears he was referring to land use changes rather than policy. Like others, Mann expressed dissatisfaction with the term safe yield since, in legal settings, it could imply that groundwater production beyond the safe yield was unsafe which "unduly dramatizes the situation." He preferred the terms sustained yield or perennial yield. Definitionally, Mann thought the vagueness in the definition of safe yield—or whatever term was used—could be justified given that it was a general definition of a term that could consider a number of conditions. Mann also noted that nonconsumptive use—produced irrigation waters that percolate back to the water table—needed to be considered where appropriate.

Mann also tackled the economic aspects of defining and quantifying safe yield by asking: "Is the amount of water available decreased by the inability of someone to justify high pumping costs?" He further mentions impediments to the use of groundwater, including a "fringe-area pumper" in a shallow part of the aquifer who could litigate to prevent deeper pumpers from producing. Mann also felt that "[i]t is hydrologically unsound to consider rising waters as normal surface flow, and yet there is a common tendency to make no legal distinction between these." Mann proposed that safe yield should only be defined by how much can be pumped rather than flows in the aquifer.

Several engineers commented on Mann's paper. Thomas (1963) said he used operating yield instead of safe yield. Suter (1963) agreed with Mann's proposal to rename safe yield and personally used perennial yield but even that had issues due to pumping distribution and the difficulty in calculating it. Suter noted that the safe yield can be calculated after an aquifer was overdrawn, but by then it was too late. In response to the comments, Mann (1963) mentioned an example where the safe yield changed in response to changing conditions and aquifer response.

Domenico and Schwartz (1990) liked Lee's (1915) definition as a philosophical concept but felt it was too ambiguous for use but then criticized the overspecification of the term by Conkling (1946) and Banks (1953).

The First 50 Years

Fifty years after Lee (1915) first introduced the concept for and name of safe yield for groundwater, the term continued to be used, a testament to its usefulness. As discussions progressed and society's needs evolved, safe yield's tent appropriately expanded from Lee's focus on preventing non-sustainable use to also include impacts to water quality, land subsidence, existing pumping rights, and economics. Several experts realized that determinations of safe yield needed to be adaptive to respond to an improved understanding of the aquifer and how groundwater development progressed.

Although there was a lot of handwringing over the ambiguity of safe yield, that ambiguity was somewhat embraced in the 1960s as a necessity—an ambiguity that continued to serve the concept well. And although many authors proposed alternative names, safe yield continued to be used as the touchstone for sustainable management. Todd (1959) produced, in my opinion, the best definition: "the amount of water which can be withdrawn from it annually without producing an undesired result." However, although Todd intended his definition to represent an amount of pumping that could be maintained indefinitely, that it is not clear from the definition alone. In some states for some basins, the safe yield has been defined through the courts and groundwater allocated accordingly (Thomas 1959; Hirshleifer et al. 1960).

Although a couple of authors brought up policy considerations with safe yield, none built policy into the definition. Parker (1951) directly discussed the issue by noting that safe yield did not consider ethical or justifiable production while Mann (1963) flirted with "cultural conditions." By the end of 1965, although many recognized the impacts of production on springs and stream flows, none considered those impacts important enough to incorporate into safe yield. Safe yield remained focused on meeting direct human needs with no consideration of environmental impacts.

In the early days of safe yield, scientists and engineers assumed that managing aquifers sustainably—somewhat like how surface-water reservoirs were managed—was, by default, the most logical and, therefore, correct way to manage groundwater. However, groundwater producers and many economists had a different perspective.

Questions

What are the pros and cons of the definition of safe yield?

Why were some scientists and engineers uncomfortable with the concept of safe yield?

How are aquifers (groundwater basins) similar to and different than surface-water basins?

Are any aquifers in your area managed toward a safe yield?

References

Allen, M., 1915, Discussion of "The determination of safe yield of underground reservoirs of the closed basin type:" Transactions of the American Society of Civil Engineers, v 78, n 1, p 239–242.

Baker, D.M., 1946, Discussion on "Utilization of groundwater storage in stream system development": Transactions of the American Society of Civil Engineers, v 111, n 1, p 343–346.

Banks, H.O., 1953, Utilization of underground storage reservoirs: Transactions of the American Society of Civil Engineers, v 118, p 220–234.

Charles, J.R., 1946, Discussion on "Utilization of groundwater storage in stream system development": Transactions of the American Society of Civil Engineers, v 111, n 1, p 319–326.

Conkling, H., 1946, Utilization of groundwater storage in stream system development: Transactions of the American Society of Civil Engineers, v 111, n 1, p 33–62.

Curtis, I., Hook, P., Morris, S., and Sumner, B., 1979, From Safety to Where… ? *in* EARCOM 2 – Contradiction, Fast Product, Edinburgh, England. [the title of this chapter is inspired by this release]

Domenico, P.A., and Schwartz, F.W., 1990, Physical and Chemical Hydrogeology: John Wiley & Sons, 824 p.

Drummond, G.B., 1946, Discussion on "Utilization of groundwater storage in stream system development": Transactions of the American Society of Civil Engineers, v 111, n 1, p 337–338.

Forbes, H., 1946, Discussion on "Utilization of groundwater storage in stream system development": Transactions of the American Society of Civil Engineers, v 111, n 1, p 346–338.

Hardin, G., 1968, The tragedy of the commons: Science, v 162, n 3859, p 1243–1248.

Hirshleifer, J., DeHaven, J.C., and Milliman, J.M., 1960, Water Supply – Economics, technology, and policy: University of Chicago Press, 1960, p 59–60. https://babel.hathitrust.org/cgi/pt?id=mdp.39015011501270&view=1up&seq=11.

Kazmann, R.G., 1951, The role of aquifers in water supply: Transactions of the American Geophysical Union, v 32, p 227–230.

Kazmann, R.G., 1956, "Safe yield" in ground-water development, reality or illusion? Journal of the Irrigation and Drainage Division, Proceedings of the American Society of Civil Engineers, Paper 1103, 12 p.

Kelley, V.A., Ewing, J., Jones, T.L., Young, S.C., Hamlin, S., Jigmond, M., Harding, J., Pinkard, J., Yan, T.T., Scanlon, B., Reedy, B., Beach, J., Davidson, T., and Laughlin, K., 2014, Updated groundwater availability model of the Northern Trinity and Woodbine aquifers: Final model report to the North Texas Groundwater Conservation District, Northern Trinity Groundwater Conservation District, Prairielands Groundwater Conservation District, and Upper Trinity Groundwater Conservation District; Volume 1, INTERA, variously paginated (990 total pages).

Lee, C.H., 1912, An intensive study of the water resources of a part of Owens Valley, California: U.S. Geological Survey Water-Supply Paper 294, 144 p.

Lee, C.H., 1915, The determination of safe yield of underground reservoirs of the closed basin type: Transactions of the American Society of Civil Engineers, v 78, n 1, p 148–151.

Leib, D.I., and Stiles, T.C., 1998, Yield estimates for surface-water sources *in* Sophocleous, M., editor, Perspectives on Sustainable Development of Water Resources in Kansas, Kansas Geological Survey Bulletin 239.

Mace, R.E., Dutton, A.R., and Nance, H.S., 1994, Water-level declines in the Woodbine, Paluxy, and Trinity aquifers of North-Central Texas: Gulf Coast Association of Geological Societies Transactions, v 44, p 413–420.

Mann, J.F., Jr., 1961, Factors affecting the safe yield of ground-water basins: Journal of the Irrigation and Drainage Division, Proceedings of the American Society of Civil Engineers, v 87, n 3, p 63–69.

Mann, J.F., Jr., 1963, Factors affecting the safe yield of ground-water basins: Transactions of the American Society of Civil Engineers, v 128, Paper No. 3434, p 180–186.

Mann, J.F., 1963, Response to comments on "Factors affecting the safe yield of ground-water basins:" Transactions of the American Society of Civil Engineers, v 128, Paper No. 3434, p 189–190.

McGuinness, C.L., 1951, The water situation in the United States with special reference to ground water: U.S. Geological Survey Circular No. 114, 127 p.

Meinzer, O.E., 1915, Discussion of "The determination of safe yield of underground reservoirs of the closed basin type:" Transactions of the American Society of Civil Engineers, v 78, n 1, p 234–239.

Meinzer, O.E., 1920, Quantitative methods of estimating ground-water supplies: Bulletin of the Geological Society of America, v 31, p 329–338.

Meinzer, O.E., 1923, Outline of ground-water hydrology – Definitions: U.S. Geological Survey Water-Supply Paper 494, 71 p.

Meinzer, O.E., 1932, Outline of Methods for Estimating Ground-Water Supplies: U.S. Geological Survey Water-Supply Paper No. 638-C, p 99–144.
Muckel, D.C., 1955, Pumping ground water so as to avoid overdraft: Yearbook of Agriculture 1955, U.S. Department of Agriculture, p 294–301.
Owen, J., 1915, Discussion of "The determination of safe yield of underground reservoirs of the closed basin type:" Transactions of the American Society of Civil Engineers, v 78, n 1, p 219–221.
Parker, G.G., 1951, Geologic and hydrologic factors in the perennial yield of the Biscayne Aquifer: Journal of the American Water Works Association, v 43, p 817–834.
Smith, G.E.P., 1915, Discussion of "The determination of safe yield of underground reservoirs of the closed basin type:" Transactions of the American Society of Civil Engineers, v 78, n 1, p 221–234.
Smith, G.E.P., 1924, The effect of transpiration of trees on the ground-water supply (unpublished paper given before the Geological Society of Washington, Nov. 22, 1922): see Journal of the Washington Academy of Sciences, v. 14, no. p. 160.
Sopp, C.W., 1946, Discussion on "Utilization of groundwater storage in stream system development:" Transactions of the American Society of Civil Engineers, v 111, n 1, p 326–330.
Stuart, W.T., 1945, Groundwater resources of the Lansing area: Michigan Geological Survey Progress Report 13, 35 p.
Suter, M., 1963, Discussion of "Factors affecting the safe yield of ground-water basins: Transactions of the American Society of Civil Engineers, v 128, Paper No. 3434, p 188–189.
Theis, C.V., 1940, The source of water derived from wells – Essential factors controlling the response of an aquifer to development: Civil Engineer, v 10, p 277–280.
Thomas, H.E., 1951, *The Conservation of Groundwater*: McGraw-Hill Book Company, New York.
Thomas, R.O., 1959, Legal aspects of ground water utilization: Proceedings of the American Society of Civil Engineers, v 85, n 4, p 41–63.
Thomas, R.O., 1963, Discussion of "Factors affecting the safe yield of ground-water basins: Transactions of the American Society of Civil Engineers, v 128, Paper No. 3434, p 186–188.
Todd, D.K., 1959, *Groundwater Hydrology*: John Wiley and Sons, Inc., New York, 336 p.

Tolman, C.F., 1937, *Ground Water*: McGraw-Hill Book Co., Inc., New York, N. Y., 1937, 593 p.
Williams, C.C., and Lohman, S.W., 1949, Geology and ground-water resources of a part of south central Kansas: Geological Survey of Kansas, Bulletin 79, 212 p. http://www.kgs.ku.edu/Publications/Bulletins/79/index.html.
Wolff, H.C., 1911, The utilization of the underflow near St. Francis, Kansas: U.S. Geological Survey Water-Supply Paper 258, p. 104–107.

5

The Paradox of Safe Yield

"Earth provides enough to satisfy everyone's need but not for everyone's greed."
Mahatma Gandhi (as quoted by Pyarelal *1956*)

A surface-water reservoir uses both inflows to the reservoir and storage in the reservoir to meet water demands over time. The safe yield of a surface-water reservoir is the consistent annual use of a reservoir such that it can provide reliable water through the drought of record (a record that has and will likely continue to change). The safe yield is expressed as a volumetric rate—such as 6800 megaliters per year, meaning that this volume is reliably available for use from the reservoir given the historical record. During an exact repeat of the drought of record (an unlikely scenario), the conservation storage of the reservoir (the volume dedicated to meeting water demands) declines to zero on the last day of the drought. Thereafter, the reservoir refills due to the end of the drought.[1]

[1] The odds that the next severe drought exactly repeats the previous drought of record are exceedingly small. Furthermore, the historical record may be small, thereby underestimating the actual drought of record. And then there's climate change, which may result in droughts worse than the drought of record.

R. E. Mace, *Groundwater Sustainability*, Palgrave Studies in Environmental Sustainability, https://doi.org/10.1007/978-3-031-13516-3_5

Thomas (1959) noted that for a surface stream it is clear when the safe yield is exceeded: the stream or reservoir is empty. Unlike a surface-water reservoir, an aquifer managed sustainably generally only relies on inflows, not storage, to meet water demands. This statement is not completely correct because, as water levels rise and decline in a sustainably managed aquifer, there are increases and decreases in storage. But these changes in storage are generally small compared to the total amount of storage in an aquifer. One example is the Edwards (Balcones Fault Zone) Aquifer in Texas. Recharge to this highly responsive, karst aquifer is, on average, 800 gigaliters per year, while the amount of water stored in the Edwards Aquifer is about 56,000 gigaliters (Maclay 1995). That means that there's about 70 years of water stored in the aquifer outside of the recharge rate. Furthermore, only 3 percent of the total aquifer storage is held between the aquifer's high stand and historical low stand when one of the major springs stops flowing (Hovorka and others 1998).

This paradox—an immense volume of water in storage compared to the volume of inflow or the allowable sustainable production—is what makes it easy for groundwater producers to produce more than sustainable amounts of groundwater. Depending on the aquifer, consequences of nonsustainable production may not be felt for decades, such as in the Ogallala Aquifer.

Groundwater Mining

There are academic and policy arguments over the ideal way to manage groundwater, and then there is what people are actually doing. In the United States, at least, by the mid-1950s, the states applied a hodgepodge of approaches, some adopting the concept of safe yield and others taking a hands-off approach to groundwater management, resulting in the tragedy of the commons discussed in Chap. 4. Bagley (1961) noted that, based on Texas and California, at least 25 percent of the groundwater produced in the United States by 1960 exceeded sustainable levels.

Conkling (1946), in his study of the alluvial aquifers of the western United States, including the Ogallala Aquifer, brought a dose of reality to the safe yield discussion in describing what was happening in practice in groundwater management. He noted that (1) groundwater use increased after droughts, (2) early considerations of aquifer production concerned how much a well could produce and if it could be produced affordably, and (3) that an aquifer wasn't considered as a reservoir to be managed until there was already a large use of water. Although Conkling did not directly address groundwater mining—the systematic depletion of water stored in an aquifer—he alluded, in the above points, to issues with groundwater management, namely that (1) individual users were more concerned about how much groundwater they could produce from their property rather than how much the aquifer could produce over a long time period and (2) discussions of the broader consequences of large volumes of unregulated pumping did not occur until there were large volumes of unregulated pumping.

McGuinness (1951) observed that there is always a reduction of storage in an aquifer (as evidenced by a decline in water levels) in response to production, including during sustainable production. For sustainable production, that reduction in storage is generally a one-time source of water as an aquifer comes into a new equilibrium. In real systems, water levels, and hence storage, may increase or decline due to seasonal changes in production and recharge. McGuinness also observed that "[i]n the mining of ground water…there is a progressive loss of storage as the withdrawal continues to exceed" what the aquifer can support in a sustainable manner. He pointed to the Southern High Plains in Texas as an extreme example of groundwater mining where production far exceeded estimates of recharge[2] as well as central and southern Arizona and California.

McGuinness (1951) proposed the concept of optimum yield: the rate at which it is found feasible and desirable to withdraw water from an underground reservoir, which could include the complete draining of an aquifer "if we decide that the best interests of an area or of the Nation as a whole are thus served." McGuinness stated that "[t]he maximum usefulness of the ground-water reservoirs over an indefinite period obviously

[2] He reported groundwater production at 20 times the rate of recharge; modern estimates of this disparity are closer to 6 times. Studies show that there is more recharge occurring than previously thought.

calls for development of their sustained yield rather than mining of their water" but noted that in cases where development had already far exceeded recharge, an alternative approach might be needed.

This recognition of production exceeding the safe yield resulted in attempts to refine the definition of "overdraft." Snyder (1955) defined five types of groundwater overdraft:

(1) developmental overdraft (as water levels decline from initial pumping),
(2) seasonal overdraft (over a pumping season),
(3) cyclical overdraft (over a drought),
(4) long-run overdraft (pumping in excess of the sustained yield resulting is long-term water-level declines), and
(5) critical overdraft (where water-level declines result in undesired results such as salt-water encroachment, land subsidence, and decreased well yields).

Snyder equated any decrease in aquifer storage, as reflected in a decrease in water levels, as an overdraft, although most today only consider production that leads to unsustainable use as an overdraft (what Snyder refers to as critical overdraft).

Thomas (1955) noted that "public opinion so far appears to favor treating ground water as a nonrenewable resource comparable to petroleum and metals, and mining it until the supply is exhausted, rather than curbing the withdrawals at an earlier date." He also noted that "[t]he water in areas of ground-water mining, however, may have great storage volume but negligible replenishment, and thus may not qualify as a renewable resource." Although Thomas attempted to justify groundwater mining from aquifers that have "negligible replenishment," in practice, "negligible" is often defined by how much is recharged compared to how much is or can be produced. In many aquifers, especially productive ones, it is relatively easy to produce more than the recharge rate even if the recharge rate is large. For example, modern estimates of recharge to the Ogallala Aquifer in Texas are about 1.2 million megaliters, a substantial amount of recharge to any aquifer.

Milliman (1956) observed that groundwater producers had "no incentive to maximize the present value of total extractions over time because [they have] no property rights which are valid in the future," a different way of defining the tragedy of the commons. In other words, in a depleting aquifer, if you don't use your groundwater rights today, they may be diminished or dissolved in the future as everyone else races to produce as much as they can. Milliman also criticized aquifer management that favored agricultural use over higher-value uses such as manufacturing. Hirshleifer and others (1960) argued that "there is no difference in economic principle between 'mining' water and mining coal or iron ore" and noted that aquifers would reach economic equilibriums based on the cost of pumping. However, Hirshleifer and others also noted that common-pool aquifers would "tend to stabilize at a level representing excessive use of the resource from the point of view of the community as a whole."

The Committee on Groundwater for the American Society of Civil Engineers (ASCE 1961) correctly observed that (1) safe yield contained a human element—what I refer to as a policy decision—and (2) safe yield, at least how it was considered up until that point, did not consider an aquifer in the broader context of the hydrologic cycle such as its impacts on surface water. They attempted to remove ambiguity from the definition of safe yield and to address a definition for groundwater mining by introducing four defined yields: maximum sustained yield, permissive sustained yield, maximum mining yield, and permissive mining yield. Maximum sustained yield is the maximum rate groundwater can be produced perennially. Permissive sustained yield is the maximum rate groundwater can be economically and legally produced without causing an undesired result. Maximum mining yield is the total volume of water that can produced, and permissive mining yield is maximum volume of water that can be economically and legally produced for beneficial purposes without causing an undesired result. The American Society of Civil Engineers used "maximum" and "permissive" such that maximum was independent of economics and policy, while permissive was dependent on it, noting that the permissive sustained yield was essentially equivalent to safe yield.

The Dismal March of the Economists

The pros and cons of draining aquifers versus managing them sustainably is an area ripe for economists. I have to confess to being suspicious of economic analyses since all externalities—the unintended costs or benefits to a third party—are not considered and, in many cases, difficult to quantify, both practically and philosophically. My suspicions were not helped when I asked an economics friend about the pros and cons of sustainability, and his response was "What do you want the answer to be?" Nevertheless, I will attempt to provide a brief overview of the economics of groundwater sustainability for consideration. There are, of course, textbooks dedicated to natural resource economics (such as Griffin 2006; Conroy and Peterson 2013; Deal 2016; Anderson 2019). My goal here is to provide enough background for readers to ask questions of economists when discussing groundwater or evaluating groundwater issues with an economics overlay.

Bagley (1961) provided a relatively simple explanation of the economics between groundwater depletion and sustainable development. I have approximately reproduced his example below but with metric instead standard units. The setup for the aquifer and economic analysis is:

- 100,000,000 megaliters of water in storage in the aquifer
- 1,000,000 megaliters per year is the safe yield for the sustainable scenario
- 5,000,000 megaliters per year is the production for the mining scenario
- 1 megaliter is worth \$1
- The discount rate is 2 percent

Net present value is a way of comparing various investment options in the same way in today's dollars where the discount rate is essentially inflation. In our present case, we have two investment options: (1) mine the aquifer to full depletion or (2) use the aquifer sustainably. Because of inflation, we all know that \$100 in cash today is likely to be worth less a year from today (e.g., the buying power of your \$100 today would effectively be \$98 in a year with a discount rate of 2 percent).

Bagley (1961) did not consider recharge in his mining calculation, but there is value in making his calculation because it places a price on the groundwater in storage (and the end result is broadly the same). Pumping 5 million megaliters a year from a storage of 100 million megaliters results in 20 years of production and a net present value of that water of about $82 million. If we indefinitely produce the aquifer sustainably at 1 million megaliters a year, the net present value is $50 million.

According to this analysis, it makes more economic sense to pump an aquifer dry in the short term than manage it for the long term. This seems, to me, counterintuitive. Yes, it will take 100 years to achieve the same total volume as the mining scenario, but after that the total volume of water produced exceeds that of the water in storage. Produce that amount for another 400 years, and sustainable development will have resulted in five times as much produced groundwater. In fact, the analysis above assumes that an infinite volume of water is produced in the sustainable scenario! However, the key phrase in this analysis is "net *present* value." With a discount rate of 2 percent, that megaliter is only worth $0.14 in 100 years, a decrease in value of 86 percent from today. If the discount rate was 4 percent, that megaliter is only worth $0.02 in 100 years, a decrease in value of 98 percent from today. The higher the discount rate, the more depletion is economically favored (Bagley 1961).

If we assume that the safe yield is equal to recharge and that the full volume of that recharge is available for production, we can produce from the aquifer for 25 years instead of 20 years resulting in a net present value of about $95 million. An interesting result of including recharge in the analysis is that annual volume of water becomes available (with some assumptions) after the aquifer is fully depleted. In this case, someone can (and does in the case of Domenico and others 1968; Domenico 1972) argue that groundwater mining is a temporary condition that allows increased production before depletion reduces production to sustainable recharge rates, although this approach ignores externalities.

Salk (1992) asked the question "Are we being good ancestors?" Net present value allows people to make economic decisions today from their (perhaps selfish) perspective. But what is that water worth in the future? Although there are a number of factors to consider, it's difficult to imagine water being worth *less* in the future, and, most likely—with increasing population and drought, will be worth more.

Bagley's simple example is useful to understand concepts and is realistic in the sense that many people make economic decisions based on benefits today rather than benefits to the future. The unsustainable development of the Ogallala Aquifer is an example of family farmers directly affecting their descendants' ability to irrigate in the future.

This analysis does not consider what economists call externalities—positive or negative consequences, intended or not, not considered in the economic analysis. Negative consequences from mining groundwater, some of which were identified by Laughlin (1971), include loss of spring and groundwater-fed stream-flows and the subsequent loss of people, businesses, and environments that rely on them. Increased costs of deepening wells for minor users, such as households, increased pumping lifts, larger users having to seek alternative water supplies. The net value described above may only relate to the farmer, but those dollars also support local businesses that support the farmer (e.g, banks, implement dealerships, services, and Dairy Queens) nor the manufacturers that rely on the materials produced by the farmers (and their associated economic impacts).

And how much is a spring really worth? Or a continuously flowing stream and river? And ecosystems and species that rely on those flows? Does everything have to have a dollar value to assess whether or not to keep it around? My team once hosted a philosopher at San Marcos Springs, and I asked him how much the springs were worth. He replied with "How much are the springs worth to that bird chirping over there? Or to that turtle in the lake?"

Renshaw (1963) concluded that economics required groundwater producers to produce more groundwater. He noted that "individuals are not encouraged to maximize the present value of total extractions over time" and introduced the concept of optimal mining, the amount of production that maximizes economic returns. Using an example from the Ogallala Aquifer in Texas, Renshaw found that mining groundwater could result in substantial economic returns.[3] Burt (1964, 1967) similarly investigated the optimal allocation of groundwater that is fixed in supply or only partially renewable.[4] Hartman (1965) reviewed the economics of groundwater development and concluded that the "interrelated nature of ground-water use and external effects resulting from uncoordinated individual action lead to inefficiencies in ground-water development and use unless appropriate institutional arrangements have been made."

In 1965, after a careful lawsuit by an irrigator, the U.S. Internal Revenue Service allowed for a groundwater depletion allowance—in other words, a tax deduction based on groundwater depletion—for producers in the Southern High Plains (essentially the Ogallala Aquifer in Texas and New Mexico south of Amarillo) (Palmer 1965; Sato 1966; Bagley 1972; Sellers 1973). Bagley (1972) noted that while the tax deduction for depletion might be expected to have an economic impact such as increasing production or the value of property, neither was seen at the time of his review. The court decision stated that it would take the aquifer more than 4000 years to refill the aquifer once emptied (although more recent estimates of recharge suggest 400 years).

Bear and Levin (1967) introduced the term optimal yield as a general term to replace safe yield and mining yield. They based the optimal yield on the physical characteristics of the aquifer, economics, and the features of the water resources system the aquifer was a part of, thus placing the yield in a broader context. To calculate the optimal yield, they propose defining an objective function, usually an economic or social goal subject to hydrological, technological, legal, political, budgetary, and other constraints, and then optimizing the system to that function to quantify the optimal yield. Depending on the objective function and the economic goals, the optimal yield could include mining.

---

[3] Using the Ogallala Aquifer in Texas as an example of an aquifer where the recharge is low compared to production, he noted that pumping was about 6.2 million megaliters per year while recharge was estimated at 37,000 megaliters per year. More recent estimates have the recharge rate to the Ogallala Aquifer in Texas at about 1.4 million megaliters per year.

[4] "Partially renewable" is not a scientific term and is loaded with potential interpretive shenanigans. Nearly all aquifers can be considered partially renewable because nearly all aquifers can be produced unsustainably. As a result, "partially renewable" is a circular argument to justify groundwater mining.

Domenico and others (1968) developed an equation to evaluate "the economic worth of groundwater mining to the remaining worth of a basin after it has been partially depleted permits establishment of an optimal, one-time storage reserve that may justifiably be exploited." They described perennial use and full depletion as "extreme conditions" and that optimal mining yield maximizes present worth, "a one-time volume of nonrenewable water that may take on values ranging from zero to the maximum amount of usable water in storage."

Mann (1968) observed that, in some areas, the maximum benefit of groundwater through mining was occurring with no plans of replacing the water once it was gone. But he also noted that, more commonly, groundwater mining developed an economy to afford the import of additional supplies to replace groundwater before it was fully depleted. Mann may have been referring to, in part, the 1968 state water plan for Texas where an ambitious scheme to import water from the Mississippi River to the High Plains to recharge and replace the depleting Ogallala Aquifer (TWDB 1968).

Young (1970) argued that the market provided "a sutiable [sic] mechanism for allocating resources" and questioned why hydrologists and engineers were offering guidance on managing groundwater instead of the market. Young recognized the connection between surface water and groundwater and that groundwater production could affect surface-water rights, but he felt that the safe yield resulted in "Thou shalt not withdraw water if it commits an externality upon they [sic] neighbor." Young thought that sustainable management ignored the time value of water and that mining groundwater was easy to justify economically. Young ultimately recommended that the maximum net economic yield—a cost-benefit analysis that provides the greatest value from a limited inventory of resources—be used to manage groundwater.

Freeze (1971) used the terms basin yield, natural basin yield, maximum basin yield, and basin safe yield. Basin yield was a general term referring to how much could be produced sustainably from a groundwater basin while natural basin yield referred to the volume of water naturally flowing through an aquifer. Maximum basin yield was the maximum amount that could be pumped sustainably without causing undesirable effects (essentially the safe yield), and basin safe yield was the maximum basin yield with a safety factor, and thus a lower value than the maximum basin yield. Freeze used a three-dimensional numerical groundwater flow model with unsaturated processes to quantify maximum basin yield.

In his textbook, *Concepts and Models in Groundwater Hydrology*, Domenico (1972) summarizes the state of yield analysis up through ASCE (1961) and stated that "[t]he question whether groundwater should be managed on a sustained- or mining-yield basis is not yet fully resolved and is controlled more by local conditions and demands than by policy decisions in advance of their absolute necessity." Domenico then considered socioeconomics and social welfare in groundwater management decisions, noting that conventional groundwater depletion analysis only focused on the costs and revenues of groundwater producers rather than a socially oriented policy that considers all costs and returns, including broader economic and social impacts. He pointed to efforts by Krutilla and Eckstein (1958), Hufschmidt (1965), and Hartman (1965) on human welfare and economic efficiency. Domenico highlighted a "rule" developed by Buchanan (1959) that "any social, legal, economic, or institutional change is desirable which results in (1) everyone's being better off or (2) someone's being better off and no one's being worse than before the change."

Domenico (1972) attempted to address the common pool problem, noting that "pumpers overlying a basin are said to have common ownership in the pool regardless of man's attempt to assign private property rights of pumpage" and that "some type of intervention is desirable when the onus of exploitation cost is nonspecifically distributed among members of the pumping community, or society as a whole." He pointed to work by Scott (1955) and Hirschleifer and others (1960) showing that centralized decision-making and quotas (i.e., permits) could be used.

Based on his review of the thinking at the time, Domenico (1972) amplified the concept of optimal yield first introduced by Bear and Levin (1967) by infusing it with more robust considerations of economic factors. Domenico included in his book ways to evaluate how many years it takes until sustainable production results in greater annual economic impact than depletion and the timing of water imports to replace depleting supplies. Ultimately, he deemed safe yield "conservative" and "idealistic," observing that it ignored the large volume of water in storage, something private landowners considered profitable to deplete. Nonetheless, Domenico concluded that, when considering the management of an aquifer in isolation, in other words, absent conjunctive use with other water sources, that "an inescapable proposition is that use rates must somehow be made to converge on safe yield."

Aiken and Supalla (1979) observed that groundwater depletion was one of the major water policy issues facing the western United States. They discussed the concept of economic depletion where the cost of producing groundwater exceeds the benefit, especially for irrigators. Aiken and Supalla reported that states had not reduced pumping through regulation due to the perception of political difficulty. Beaumont (1985) reported that he was seeing a decline in irrigated acreage and an increase in the number of wells to irrigate the same dwindling number of hectares in the Ogallala Aquifer and stated that "the next fifty years may well follow its predecessor, the Bison herds, if not into near extinction, then at least to a much lower level of activity."

Price (2002), in a punchy defense of groundwater depletion, argued that the unsustainable pumping of the Chalk Aquifer in England allowed London to develop into a population and manufacturing center (Downing 1986, 1993).[5] Price observed that most fresh water is stored in aquifers and that it "is inconceivable that we should not make full use of it." Price speculated that the reason sustainable groundwater management is desired is to preserve the "sublime phenomena" of springs, wetlands, perennial streams, and the associated ecosystems. When considering sustainable management, Price suggested that three factors needed to be considered: (1) what is being preserved and why, (2) how to preserve it, and (3) that most things in life are a compromise. Price stated that reducing pumping to improve a private amenity is usually not worth it.

Griffin (2006) criticized safe and sustainable yield concepts because "these criteria make no attempt to weigh the human benefits received from the water against the losses from the deterioration of the aquifer and/or the environment." In other words, decisions on groundwater management should be purely economic decisions. Griffin (2017) stated that sustainable yield is "useless as a policy guide" because it did not account for the loss of human benefits due to reduced pumping.

[5] A similar argument could be made for San Antonio after local production from the Edwards Aquifer causes San Antonio and San Pedro Springs to become intermittent.

The Bottom Line

Perhaps a key issue with economic analysis is discounting the value of future benefits *a la* the Bagley (1961) example near the beginning of this chapter, something, as Griffin (2017) points out, is accepted economic doctrine. But does that really work with water and future generations? In the example at the beginning of the chapter, the value of the water in 100 years is discounted 86 percent with a 2 percent discount rate. But what is the value of that water in 100 years to the people living 100 years from now? It almost certainly is not worth less and, more than likely, is worth more, perhaps much more. As a Congressional Research Service report noted, there is disagreement on whether the discount rate accounts for "society's preference for consumption today versus the future" (Redacted and Redacted 2016) or the appropriateness of today's societal valuing (or discounting) benefits for future society (McCarthy and others 2001, p. 97). In short, traditional economic analysis favors today's generation over tomorrow's.

Furthermore, not everything in life is a choice between which scenario has the largest economic outcome. The magazine *The Economist* famously refers to economics as "the dismal science," a phrase first coined by Thomas Carlyle while making an economic argument for the reintroduction of slavery to the West Indies (Carlyle 1849). Economic intangibles concerning culture and place also have a role in making decisions. Just as science provides information for policymakers to consider, so does economics and other concerns.

Questions

Why is safe yield a paradox?
What are the pros and cons of producing groundwater in storage?
Should we care about people in the future?
What else is affected when groundwater is not produced sustainably?
How much is a spring worth?

References

Aiken, J.D., and Supalla, R.J., 1979, Ground water mining and western water rights law – The Nebraska experience: South Dakota Law Review, v 24, p. 607–648.

Anderson, D.A., 2019, Environmental Economics and Natural Resource Management, 5th Edition: Routledge, New York, 446 p.

ASCE (American Society of Civil Engineering), 1961, Ground water basin management: Manual of Engineering Practice No. 40, 160 p. https://babel.hathitrust.org/cgi/pt?id=coo.31924000122311&view=1up&seq=8.

Bagley, E.S., 1961, Water rights, law and public policy relating to ground water "mining" in the Southwestern States: The Journal of Law & Economics, v. 4, p. 144–174.

Bagley, E.S., 1972, The Ground-Water Depletion Allowance under the Federal Income Tax: Natural Resources Journal, v 12, n 3, p 445–466.

Bear, J., and Levin, O., 1967, The optimal yield of an aquifer: International Association of Scientific Hydrology Bulletin, Publication 72, p. 401–412.

Beaumont , P., 1985, Irrigated agriculture and ground-water mining on the High Plains of Texas, USA: Environmental Conservation, v 12, n 2, p 119–130.

Buchanan, J.M., 1959, Positive economics, welfare economics, and political economy: Journal of Law and Economics, v. 2, n. 2, p. 124–138.

Burt, O.R., 1964, Optimal resource use over time with an application to ground water, Management Science, v. 11, p .80–93.

Burt, O.R., 1967, Temporal allocation of groundwater, Water Resources Research, v. 3, no. 1, p. 45–56.

Carlyle, T., 1849, Occasional discourse on the Negro question: Fraser's Magazine for Town and Country, v 40, p 670–679.

Conkling, H., 1946, Utilization of groundwater storage in stream system development: Transactions of the American Society of Civil Engineers, v 111, n 1, p 33–62.

Conroy, M.J., and Peterson, J.T., 2013, Decision Making in Natural Resource Management – A Structured, Adaptive Approach: Wiley-Blackwell, New York, 480 p.

Deal, K.H., 2016, Wildlife and Natural Resource Management, 4th Edition: Delmar Cengage Learning, Boston, 340 p.

Domenico, P.A., 1972, *Concepts and Models in Groundwater Hydrology*: McGraw-Hill, New York, 405 p.

Domenico, P.A., Anderson, D., and Case, C., 1968, Optimal ground-water mining: Water Resources Research, v 4, n 2, p 247–255.

Downing, R.A., 1986, Development of groundwater *in* Brandon, T.W., editor, *Groundwater Occurrence, Development and Protection*: Institution of Water Engineers and Scientists, London, p 485–542.

Downing, R.A., 1993, Groundwater resources, their development and management in the UK – An historical perspective: Quarterly Journal of Engineering Geology, v 26, p 335–358.

Freeze, R.A., 1971, Three-dimensional, transient, saturated-unsaturated flow in a groundwater basin: Water Resources Research, v. 7, n. 2, p. 347–366.

Griffin, R.C., 2006, Water resource economics – The analysis of scarcity, policies, and projects: MIT Press, Cambridge, Massachusetts, 496 p.

Griffin, J.M., 2017, Interjecting economics into the groundwater policy dialogue: Texas Water Journal, v. 8, n. 1, p. 97–112.

Hartman, L.M., 1965, Economics and groundwater development: Ground Water, v 2, n 2, p 4–8.

Hirshleifer, J., DeHaven, J.C., and Milliman, J.M., 1960, Water Supply – Economics, technology, and policy: University of Chicago Press, p 59–60. https://babel.hathitrust.org/cgi/pt?id=mdp.39015011501270&view=1up&seq=11.

Hovorka, S.D., Mace, R.E., and Collins, E.W., 1998, Permeability structure of the Edwards Aquifer, South Texas – Implications for aquifer management: Bureau of Economic Geology, The University of Texas at Austin, Report of Investigations No. 250, 55 p.

Hufschmidt, M., 1965, The methodology of water resource system design *in* Burton, I., and Kates, R., editors, *Readings in Resource Management and Conservation*: University of Chicago Press, p 558–570.

Krutilla, J., and Eckstein, O., 1958, *Multiple Purpose River Development*: The Johns Hopkins Press, Baltimore, 336 p.

Laughlin, B.N., 1971, State management of ground water mining – A system based on policy, fact and theory – Fundingsland v. Colorado Ground Water Commission: Land and Water Law Review, v 6, n 2, p 569–576.

Maclay, R.W., 1995, Geology and hydrology of the Edwards Aquifer in the San Antonio Area, Texas: U.S. Geological Survey, Water-Resources Investigations Report 95-4186, 64 p.

Mann, J.T., Jr., 1968, Concepts in ground water management: Journal of the American Water Works Association, v 60, n 12, p 1336–1344.

McCarthy, J.J., Canziani, O.F., Leary, N.A., Dokken, D.J., and White, K.S., editors, 2001, Climate Change 2001 – Impacts, Adaptation, and Vulnerability: Cambridge University Press, 1032 p.

McGuinness, C.L., 1951, The water situation in the United States with special reference to ground water: U.S. Geological Survey Circular No. 114, 127 p.
Milliman, J.W., 1956, Commonality, the price system, and use of water supplies: Southern Economic Journal, v 22, n 4, p 426–437.
Palmer, R.L., 1965, Ground water – Depletion of a wasting asset: Southwestern Law Journal, December, v 19, p 791–800.
Price, M., 2002, Who needs sustainability? *in* Hiscock, K.M., Rivett, M.O., and Davison, R.M., editors, *Sustainable Groundwater Development*: Geological Society, London, Special Publications, 193, p 75–82.
Pyarelal, N., 1956, Mahatma Gandhi – The Last Phase: Navajivan Publishing House, India, 750 p.
Redacted and Redacted, 2016, Discount rates in the economic evaluation of U.S. Army Corps of Engineers Projects: CRS Report prepared for members and committees of Congress, Congressional Research Service, R44594, 37 p.
Renshaw, E.F. 1963, The management of ground water reservoirs, Journal of Farm Economics, v 45, 285–295.
Salk, J., 1992, Are We Being Good Ancestors? World Affairs: The Journal of International Issues, v. 1, n. 2, p. 16–18.
Sato, S., 1966, Ground water rights and depletion deduction: Natural Resources Journal, v 6, n 2, p 237–247.
Scott, A., 1955, *Natural Resources – The Economics of Conservation*: University of Toronto Press, Toronto, 184 p.
Sellers, J.H., 1973, Tax implications of ground-water depletion: Ground Water, v 11, n 6, p 27–35.
Snyder, J.H., 1955, Ground water in California – The experience in Antelope Valley: Giannini Foundation Ground Water Studies No. 2, Agricultural Experiment Station, University of California, 171 p.
Thomas, H.E., 1955, Water rights in areas of ground-water mining: U.S. Geological Survey Circular 347, 16 p.
Thomas, R.O., 1959, Legal aspects of ground water utilization: Proceedings of the American Society of Civil Engineers, v 85, n 4, p 41–63.
TWDB (Texas Water Development Board), 1968, The Texas water plan: Texas Water Development Board, Austin, Texas; variously paginated (252 total pages).
Young, R.A., 1970, Safe yield of aquifers – An economic reformulation: Proceedings of the American Society of Civil Engineers, Journal of the Irrigation and drainage Division, v. 96, n. 4, p. 377–385.

6

The Rising Waters of Groundwater Sustainability

"Human beings, in their quest for economic development and enjoyment of the riches of nature, must come to terms with the reality of resource limitation and the carrying capacities of ecosystems, and must take account of the needs of future generations."
Mohamed Kassas, Mostafa K. Tolba, and John H. Loudon (IUCN *1980*)

From the earliest days, engineers and scientists thought that groundwater should be managed sustainably, similar to how engineers managed surface-water reservoirs. While some chose to manage according to safe yield concepts, other chose to not manage at all, leading to the systematic depletion of certain aquifers across the planet. On reflection, engineers, scientists, and economists justified systematic depletion with an aquifer's water budget (relatively low recharge rates) as well as for economic reasons, although economic outcomes seemed to depend on the eye of the beholder.

Although safe yield continued to be used—despite attempts to rename or reconceptualize it—it had, up until the 1970s, been focused on maximizing groundwater production to the benefit of people using an aquifer. Even then, by 1970, several authors recognized that maximizing

R. E. Mace, *Groundwater Sustainability*, Palgrave Studies in Environmental Sustainability, https://doi.org/10.1007/978-3-031-13516-3_6

sustainable groundwater production in an aquifer might affect those in adjacent aquifers and that intercepting rising waters (a poetic but no longer used term used by Conkling 1946, Thomas 1959; and Mann 1961, 1963, 1968, among others) such as springs and seeps affected surface-water rights. With the advent of environmentalism in the 1960s and 1970s, rising waters that fed spring, stream, and wetland ecosystems became a concern in and of itself and was added to the versatile definition of safe yield as yet another undesirable outcome, provided decisionmakers saw it as such.

The concept of groundwater sustainability arrived on the heels of the environmental movement in the 1960s and 1970s through broader, global discussions of sustainable development as the world's population grew along with human impacts to the planet. The International Union for Conservation of Nature and Natural Resources introduced the concept of sustainable development when discussing the long-term management of living resources such as ecosystems, genetic diversity, and sustainable use of species such as fish and forests (IUCN 1980). They concluded that "[f]or development to be sustainable it must take account of social and ecological factors, as well as economic ones; of the living and non-living resource base; and of the long term as well as the short-term advantages and disadvantages of alternative actions." The Union mentioned water with respect to ecosystem services for filtration. They recommended comprehensive management of water, including groundwater. Brown (1982), in his book *Building a Sustainable Society*, mentioned groundwater depletion as a concern for the sustainability of global food supplies.

The Brundtland Commission popularized sustainable development, defining it as "development that meets the needs of the present without compromising the ability of future generations to meet their own needs" (WCED 1987). Embedded in this definition were three pillars—environment, economics, and social equity—that needed to be considered when evaluating sustainable development (Brinkmann 2020). Social equity applies to future generations as well as the current population. Although the Brundtland Commission did not dwell on groundwater issues, they did identify impacts to water quality and depletion as concerns for groundwater (WCED 1987).

Safe Yield and the Environment

Fetter (1972) was the first to observe that "traditional factors used to determine safe yield of a groundwater basin (water supply, economics, water quality and water rights) do not include environmental effects." Others had long identified the impacts of groundwater production on surface water, but none had expressed concern for the environmental consequences. Fetter's study was focused on the connection of groundwater to a coastal estuary and proposed that environmental effects should be accounted for when determining the safe yield of source aquifers. Fetter's use of the word "traditional" and continued use of the term "safe yield" showed that he understood that safe yield was adaptable enough to embrace an evolving appreciation of factors as society and science changed with time.

Fetter (1976) estimated a safe yield for the South Fork of Long Island, New York, noting that groundwater production may drain streams and ponds and impact the environment. As an interesting aside, Anderson and Berkebile (1977) scolded Fetter for using the term safe yield, calling it "disconcerting" and pointed to criticisms by Thomas (1951) and Kazmann (1956). They agreed with Kazmann that the term should not be used and that perhaps the maximum stable basin yield proposed by Freeze (1971) was "a step toward formulating a more rational approach." Fetter (1977) responded that safe yield was a useful concept if properly employed. He also noted that "environmental impacts include ecological, economical, social, cultural, and political values" and observed that "[t]he safe yield of an aquifer system is not a physical phenomena but is, rather, a subjective phenomena based upon human values."

Freeze and Cherry (1979, p. 2) noted that if their textbook had been written a decade before, it would have been almost entirely focused on groundwater as a resource for people but acknowledged that groundwater was more than a resource—it was an important part of the natural environment. Freeze and Cherry (1979, p. 305) used the term *aquifer yield* instead of safe yield and defined it as the maximum amount of production that could be sustained by an aquifer without causing unacceptable declines in water levels. They also presented *basin yield* and defined it as the maximum amount of production that could be sustained by the complete hydrogeologic system in a groundwater basin without causing unacceptable declines in water levels or causing unacceptable changes to any other component of the hydrologic cycle in the basin.

Fetter (1980), in his textbook *Applied Hydrogeology*, presented a composite, if not unwieldy, definition of safe yield as "the amount of naturally occurring groundwater that can be withdrawn from an aquifer on a sustained basis, economically and legally, without impairing the native groundwater quality or creating an undesirable effect such as environmental damage." Building on his earlier work on the environmental impacts of groundwater production (Fetter 1972), Fetter (1980) emphasized how groundwater production might reduce streamflow, lower lakes levels, and dry wetlands and noted that "environmental impacts include ecological, economical, social, cultural, and political values." Fetter (1980) made the argument that, despite the vagueness of safe yield that gave many professionals pause, "the basic concept must be applied whenever the use of an aquifer is planned and managed." Perhaps this last point is why he named the section "Paradox of Safe Yield."

Sustainable development shifted to include the community in development discussions and decisions. The United Nations Conference on Environment and Development—Rio 1992—resulted in Agenda 21 (United Nations 1992) and outlined "an action plan for sustainable development that integrates environment and development concerns and which is strongly oriented towards bottom-up participation and community-based approaches" (Hiscock and others 2002). Downing (1993) stated that "[d]evelopment must consider all demands on an aquifer including not only water supply but environmental aspects such as maintaining river flows and wetland environments."

The Shift to Sustainability

Although sometimes presented as different than safe yield (typically because of incorrectly associating safe yield with being equal to recharge), sustainable yield is simply a different term meaning—ultimately—the same thing. As previously noted, McGuinness (1951) was the first to match the word "sustain" with the word "yield" when he used sustained yield as an alternative name for safe yield (and restricted it to a function of the properties of the aquifer as constrained by how groundwater is developed). Suter and others (1959) used the term "practical sustained yield," but did not define it.

"Sustain" teamed with "yield" next appeared in 1961 when the American Society of Civil Engineers attempted to clarify definitions of aquifer yields by introducing the terms maximum sustained yield and permissive sustained yield (CG-ASCE 1961). They defined maximum sustained yield as the maximum rate groundwater can be produced perennially and permissive sustained yield as the maximum rate groundwater can be economically and legally without causing an undesired result. Mann (1963) proposed using sustained yield or perennial yield instead of safe yield, and Walton (1964) proposed the term practical sustained yield. Walton (1964) defined practical sustained yield as "[t]he maximum amount of water that can be continually withdrawn from existing pumping centers without creating critical water levels or exceeding recharge."

Todd and Meyer (1970, 1971) appear to be the first to use the term sustainable yield for groundwater as an alternative to safe yield, a term later used by Toth (1973), Hopkins (1987), Hamlin and Anthony (1987), and others. In 1977, the American Society for Civil Engineers released "Policy Statement 243—Groundwater Management" (ASCE 1977) which called for, among other items, management of groundwater resources and protection of groundwater from depletion and degradation. The policy statement had been reaffirmed as recently as 2019 by the Board of Direction. ASCE (1998) defined sustainable water resource systems as those designed and managed to fully contribute to the objectives of society, now and in the future, while maintaining their ecological, environmental, and hydrological integrity.

Alley and others (1999) eloquently defined groundwater sustainability as the "development and use of ground water in a manner that can be maintained for an indefinite time without causing unacceptable environmental, economic, or social consequences." Alley and others (1999) stated that "[p]erhaps the most important attribute of the concept of ground-water sustainability is that it fosters a long-term perspective to management of ground-water resources." Alley and others listed four reasons groundwater management benefits from a long-term, sustainability perspective: (1) groundwater is renewable but may renew at a smaller rate than groundwater production, (2) groundwater development may take place over many years, (3) the full effects of groundwater production may not be evident for years or decades, and (4) losses from groundwater storage need to be considered over the time needed to achieve sustainability. Alley and others (1999) noted that safe yield was commonly used to quantify sustainable groundwater development and that it should be used with respect to the effects on water-levels, streamflows, and water quality. They suggested evaluating aquifer impacts with increasing use, suggesting an adaptive management approach to evaluate estimates of the safe yield.

Devlin and Sophocleous (2004) made the argument that sustainability and *sustainable production* were two different concepts that should not be used interchangeably. Sustainable production is an amount of production from an aquifer that can be maintained indefinitely while sustainability includes policy decisions to help define a sustainable production that does not produce undesired effects. I wholeheartedly endorse this argument and have tried to adhere to this in this book and in practice.

Safe yield ultimately led to sustainable yield (Alley and Leake 2004; Maimone 2004). After nearly a century of numerous attempts to rename safe yield, sustainable yield has achieved the greatest success, probably because of its connection to the broader, global discussion of sustainability or perhaps because it happens to encompass the current understanding and policy preference of unacceptable consequences. Regardless, the term "safe yield" continues to be used in literature, law, and practice, perhaps because it appears in law in many places. In these cases, it is important to review how safe yield is defined whether by law, lawyer, engineer, or hydrogeologist. Does it represent the latest understanding of unacceptable consequences, or does it represent the unacceptable consequences of a past decade, or worse (equated to recharge)?

Some hydrogeologists consider safe yield and sustainable yield to be two different concepts. For example, Sophocleous (1997) suggested that a review of hydrologic principles such as that in Theis (1940) informs the reader of why safe yield is not sustainable yield, a sentiment repeated by Zhou (2009). However, Sophocleous erroneously equated safe yield with natural (predevelopment) recharge rather than the outcome-based definition more appropriately used by Lee (1915) and others thereafter. While many, including Lee (1915) and Meinzer (1920), committed the hydrogeologic sin of equating safe yield to recharge, whether natural or natural and induced, outcome-based definitions of safe yield adapted well to changing values and mores, including the consideration of environmental and societal impacts. Later in the same article, Sophocleous appears to admit as much when he wrote that a "better definition of [safe yield] would address the sustainability of the system."

Zhou (2009) differentiates sustainable yield from safe yield by concluding that sustainable yield considers the dynamics of groundwater systems and links groundwater systems with ecological and human systems. However, safe yield's evolution grew to include aquifer dynamics and interactions with ecology and people, even before sustainable yield became the preferred term among certain scientists and practitioners.

Sophocleous (1997) was certainly correct that misinformed government officials had misused the concept of safe yield resulting in "continued ground-water depletion, stream dewatering, and loss of wetland and riparian ecosystems" and that "a breakdown in communication between scientists/engineers and water managers" had occurred, although some misinformed scientists and engineers had informed the government officials. I think part of the misunderstanding is that it is easy to think about an aquifer as a checking account where recharge is your paycheck; pumping wells, springs, and baseflow are withdrawals; and the amount of water in storage is the balance. While some aquifers work in this simple way, the vast majority are far more complicated.[1]

Alley and Leake (2004) discussed weak and strong sustainability, where *weak sustainability* assumes that the generational hand-off will include new and more efficient technology to use water and *strong sustainability* where there is no substitute for the water. However, in my opinion, it is far better to respond to actual technological advancements as they occur

[1] One issue is that some of the withdrawals—springs and baseflows—generally have to have a minimum amount in the account (minimum amount of storage) to even exist.

rather than anticipating them to occur at some time in the future.[2] It is far easier politically to allow more production from an aquifer than it is to reduce it (although there are exceptions).

How Long Is Forever?

Worster (1993), speaking generally of sustainable development, observed that the time frame for sustainability was rarely addressed. In other words, "Is a sustainable society one that endures for a decade, a human lifetime, or a thousand years?" Worster concluded that "no one really expects sustainable to mean 'forever'; that would be a utopian expectation that no society has ever achieved." Worster also identified three flaws in sustainable development: (1) "it is based on the view that the natural world exists primarily to serve the material demands of the human species", (2) "depends on the assumption that we can easily determine the carrying capacity of local regional ecosystems", and (3) "the sustainability ideal rests on an uncritical, unexamined acceptance of the traditional world view of progressive, secular materialism." I'm not sure what that last statement means, but it sounds serious.

Hisock and others (2002) argued that sustainability should be defined over a certain time period and that other sources of water should also be considered. Price (2002) stated that, when talking of sustainable use, "it is necessary to stipulate the period over which the use is planned or implemented." Evans and Cook (2002) proposed a definition for sustainable yield of "[t]he groundwater abstraction regime, measured over a specified time frame, that allows acceptable levels of stress and protects the higher value uses that have a dependency on the water" (as quoted by Kalf and Woolley 2005). The Australian Groundwater Committee included a "specific planning time frame" in its definition of sustainable yield (NGC 2004).

Gleeson and others (2010, 2012) proposed that sustainability goals for water quality and quantity should be set on a multigenerational time horizon of 50 to 100 years, while acknowledging even longer-term

[2] After all, they promised us jetpacks…

impacts." Gleeson and others (2012) suggested that mean residence time—the average time for groundwater to flow from recharge to discharge areas—be used to indicate appropriate planning horizons. They also proposed backcasting (identifying the goal and then identifying management options to achieve the goal) and managing adaptively.

I can easily disprove Worster's "no one" statement by simply disagreeing with him (which I do). Sustainable development is like being pregnant: either you are or you aren't. Sustainability defined over a certain time period can easily hide planned depletion of groundwater resources behind the word sustainability. Production may be sustainable over a specified time period, but the production may not be sustainable for the aquifer. For example, Kalf and Woolley (2005) note that a "groundwater withdrawal regime that is sustainable for 30 years, for instance, may lead to failure in year 31." Mace (2021) showed that although 87 percent of aquifer-county-groundwater management area splits in Texas[3] projected sustainable *production* over the 50-year planning period, 76 percent of the aquifers will not be produced sustainably. What is important is that the process of defining sustainability and sustainable yield is adaptive over time to changing social mores, science, development, climate, and other factors. Sustainability today may not be sustainability tomorrow.

Considering other sources of water when evaluating groundwater sustainability as suggested by Hisock and others (2002) is not groundwater sustainability; it is water-supply sustainability. I am not saying that is wrong (there is no right or wrong policy decision from a science perspective) or a bad idea; but I am saying that the term groundwater sustainability places a sustainability box around an aquifer not an aquifer and other resources. An aquifer is either produced sustainably or it is not. Over the long-term, groundwater might be strategically considered a bridging supply, a supply that is used unsustainably until another supply—such as seawater desalination—becomes available either physically or economically. But again, this is water-supply sustainability, not groundwater sustainability.

[3] "Aquifer-county-groundwater management area splits" refers to spatial areas over which the Texas Water Development Board reports how much groundwater can be produced according to goals identified by local regulating bodies.

Although groundwater sustainability should be forever, it is important to be adaptive and revisit aquifer conditions, science, and policy over time, preferably on a schedule. Domenico and Schwartz (1990) noted that in practice safe yield has no unique or constant value because of the location of production wells and how those wells interact with the hydrologic system. Depending on how groundwater production develops over time, previous estimates of sustainable yield may need to change to reflect actual production distribution. Furthermore, an estimate of sustainable yield has its own errors, especially if the aquifer has never been in the state required to achieve production levels set by the sustainable yield. Finally, changed conditions may impact the sustainable yield. For example, if urban interests buy out agriculture users and export the water, groundwater production will occur differently over the years and thus may affect estimates of sustainable yield. Climate change may impact recharge rates, thus affecting the sustainable yield. For example, Loaiciga (2016) concluded that climatic variability requires an adaptive process for evaluating a long-term sustainable yield.

Alley and Leake (2004) noted that "values relating to the sustainability of ground water resources change with time." Because of this change, stakeholder and decisionmaker assessments of groundwater sustainability should undergo systematic review. These value changes may be due to a desire to be more or less protective of springflows or, depending on the overarching regulatory scheme, may desire to abandon sustainability to meet short-term or long-term drought needs.

The Water Budget Myth

Bredehoeft and others (1982) introduced the concept of the Water Budget Myth, stating that "[p]erhaps the most common misconception in groundwater hydrology is that a water budget of an area determines the magnitude of possible groundwater development." Each aquifer has a water budget, namely an amount of water entering (recharge, cross-formational flow), an amount of water leaving (natural discharge such as springs and stream baseflow, production), and changes in the amount of water stored in the aquifer (as reflected by water levels). Hydrogeologists can estimate these flows, but generally not with great accuracy.

The myth of the water budget myth is that once you know the water budget, you can estimate how much can be produced sustainably, namely through the recharge rate or total inflows. The myth probably started at the beginning with the Lee (1915) paper. In Lee's defense, his basin aquifer probably comes the closest to a bathtub model for an aquifer if leakage from the basin is minimal. But most aquifers do not approximate a bathtub. One issue is time. Because aquifers have a resistance to flow as opposed to a bathtub, where there is no resistance, it takes time for production to come into equilibrium in an aquifer. Depending on the aquifer, achieving equilibrium can take a long amount of time (Bredehoeft and others 1982). Even in a basin aquifer like the one Lee (1915) considered, it can be between 400 and 1000 years before an aquifer is brought to a new equilibrium (Bredehoeft 2002). Confined aquifers can take even longer to reach equilibrium. Springs and streams often go dry long before production reaches the magnitude of recharge (Bredehoeft 1997).

Another issue is that the flow system in an aquifer is often leaky and far more complicated than a simple bathtub water-budget model. The sustainable yield is often considerably less than recharge if stakeholders and decisionmakers want to retain flows to springs and streams (Sophocleous 2000). And as Tolman (1937) noted, in some confined systems, sustainable yield is controlled more by aquifer properties than the dynamics of recharge and discharge.

Bredehoeft revisited the Water Budget Myth in 1997 and 2002 (Bredehoeft 1997, 2002). In his 2002 paper, he introduced a new controversy: that sustainability was independent of recharge and was solely dependent on the capture of discharge. This is perhaps too strong of a statement. As Zhou (2009) says, "[i]t is obvious that natural recharge is a very important factor in determining the sustainable pumping rate." If there's no recharge, there's no sustainable pumping. What Bredehoeft (2002) was trying to say was that the act of producing groundwater doesn't capture recharge to meet the volume of that production—it captures discharge. To capture recharge means intercepting water just as it crosses the water table. However, once water enters the aquifer, it follows a flowpath to discharge. In short, you say po-tay-toe, I say po-tah-toe—you say captured discharge, I say captured recharge. Bredehoeft (2002) noted that natural recharge is often fixed by rainfall, vegetation, and soil properties: groundwater production generally doesn't affect recharge, but it does affect natural discharge.

The Unbearable Ambiguity of Sustainability

Inevitably, there was a backlash against sustainable yield just as there was a backlash against safe yield. As Sophocleous (2000) stated, "sustainability is an idea that is broadly used but perhaps not well understood." In an editorial in the academic journal *Ground Water*, Wood (2001)—perhaps unknowingly—echoed early complaints against safe yield when he expressed dislike for sustainability because it was "value laden." He proposed the term *renewability*, defining it as "the natural average amount of water that enters the watershed each year." This definition, of course, crashes right into Bredehoeft's Water Budget Myth, but at least succeeds in developing a number that is divorced from value judgements.

Wood (2001) concludes that "[f]or scientists and engineers to maintain credibility in the formulation of public policies ... we must clearly separate the science and engineering inherent in the study of renewability from the management inherent in sustainability." Wood's concerns over maintaining a clear separation between science and policy are legit, but his solution completely divorces science from sustainability. A better approach is for the scientists and engineers to clearly understand where the line is between science and policy and only provide technical support to policymakers.

Kalf and Woolley (2005) step up to the plate to be the Kazmann of groundwater sustainability. They stated that "current definitions and concepts [of sustainable yield] are ambiguous and non-physically based so cannot be used for quantitative application" and argue that not all groundwater systems can be managed sustainably. However, Kalf and Woolley missed that management constructs are, by definition, policy decisions that require the involvement of people and preferences and that processes based on people and preferences can be (and have been) quantified. Furthermore, all aquifers can be managed sustainably; sustainable management is a question of political will rather than science. Kalf and Woolley stated that "in many semi-arid and arid environments groundwater systems cannot be sensibly developed using a sustained yield policy particularly where ecological constraints are applied" thus expressing personal policy preferences on groundwater management rather than placing those decisions where they belong: with policymakers. Scientists and engineers need to ensure that they do not impart their own policy preferences onto stakeholders and policymakers.

Even the decision to manage sustainably is a policy decision. As previously discussed, managing to a firm or safe yield in a surface-water reservoir makes perfect sense. The same is probably true in any highly responsive aquifer, an aquifer that responds quickly—on the order of days to months to less than a decade—to changes in recharge and production for the same reasons as a surface-water reservoir: overuse threatens everyone's use. However, in a slowly responsive aquifer—an aquifer that does not respond quickly to changes in recharge and production, on the order of decades or more—the impacts of nonsustainable use are greatly delayed while there is great economic benefit in the meantime. In fact, some have argued that nonsustainable use could be (and has been) an overall benefit to society (Downing 1993; Hiscock and others 2002).

Loucks (2000) noted the need to understand the hydrologic system and then let the political process make choices when there are conflicts. Loucks also recognized the need to develop systems that adapt to changes in management goals and changes to the availability of water over time. Loucks concluded that "[b]ecause sustainability is a function of various economic, environmental, ecological, social, and physical goals and objectives, water resources management must inevitably involve multi-objective tradeoffs in a multidisciplinary and multi-participatory decision-making process." Peralta and others (2011) used a stakeholder and optimization process to determine how to spatially distribute groundwater production. Pierce and others (2012) placed sustainable yield in a broader context of a consensus yield defined by stakeholders informed by science that could include sustainability or planned depletion. Sophocleous (1997) noted the uncertainty in our understanding of aquifers but suggested that we needed to make management decisions on the best available science and be flexible as new information and understanding is developed.

Gleeson and others (2012) considered groundwater sustainability "a value-driven process of intra- and intergenerational equity that balances the environment, society, and economy." They also concluded that safe yield often focused on aquifers as value-neutral physical systems; however, as we have seen, that has not been accurate since the beginning with value-laden terms such as "dangerous," "practicable," "impairing," and other value-laden terms.

Rudestam and Langridge (2014) investigated how scientists, courts, agencies, and the public define groundwater sustainability differently. In interviews with managers of local water agencies, they found ambiguity, greenwashing (sustainable development may still lead to undesired effects for some), relative timeframe, oversimplification (impression by stakeholders that there is one number), lack of transparency on how infrastructure fits in, ecological concerns, and extreme events as shortcomings to sustainable yield. One manager argued for using overdraft instead of sustainable yield since overdraft was purely a technical determination. Rudestam and Langridge (2014) found that agency managers and hydrogeologists "worried that [sustainable yield's] subjective nature and implied normative values confused the public and might lead to future conflicts or debates." This is, of course, an opportunity to educate the public as well as policymakers to increase their awareness of what groundwater sustainability is and how science and policy define it.

Rudestam and Langridge (2014) themselves conclude that sustainable yield "is ambiguous and may be interpreted to support different socio-economic and political interests" and propose that "any definition make explicit the assumptions embedded in the use of these terms in groundwater management practices."

Scientists drafted a call to action at a 2019 American Geophysical Union Chapman Conference in Valencia, Spain, that called for (1) putting a spotlight on global groundwater sustainability, (2) managing and governing groundwater sustainably, and (3) investing in groundwater governance and management (Gleeson and others 2019). As of late 2021, the statement had 1348 signatories from 102 countries. The "Putting a spotlight" call to action called for completing a United Nations World

Water Development Report, planning a global groundwater summit, and "recognizing the global importance of groundwater in the UN's Sustainable Development Goals." For management and governance, the call to action recommended "maintaining long-term, dynamically stable storage of high-quality groundwater using inclusive, equitable and long-term governance and management." Investment in groundwater governance and management involved

> "1) establishing an open, transparent global monitoring and reporting system on aquifer conditions and groundwater use;
> 2) incorporating nature-based solutions and measures for protected and managed aquifer recharge and groundwater quality protection;
> 3) coordinating management of groundwater, surface water and new water sources;
> 4) incorporating groundwater into all SDG [sustainable development goals] pathways related to water, energy and food security;
> 5) changing the economic incentives of direct and indirect groundwater users;
> 6) strengthening inclusive groundwater management bodies at all levels;
> 7) developing institutional groundwater management capacity at all levels and enhancing global networking and sharing of best practices;
> 8) raising awareness with stakeholders representing agricultural, industrial, urban, domestic and environmental water users; and
> 9) holding global, high-level deliberations on groundwater management, including transboundary basins and aquifers."

Gleeson and others (2020) proposed a definition of groundwater sustainability as "maintaining long-term, dynamically stable storage and flows of high-quality groundwater using inclusive, equitable, and long-term governance and management." Whereas the Alley and others' (1999) definition was agnostic on how unacceptable environmental, economic, and social consequences are determined, Gleeson and others (2020) added a requirement of inclusivity, equity, and long-term governance (I discuss groundwater governance in Chap. 7). In other words, Gleeson and others (2020) have added specificity on *how* groundwater sustainability is determined and subsequently governed in the future to the definition of groundwater sustainability.

While I think Gleeson and others' (2020) goals are noble and desirable, I also think they are too restrictive. For example, what if a benevolent autocrat decrees that groundwater shall be produced in a manner to (1) not reduce springflows more than 10 percent of preproduction levels and (2) not reduce the productivity of shallow wells owned by the poor more than 15 percent of their productivity? Under the Gleeson and others' (2020) definition, that would not be groundwater sustainability although it would be under the Alley and others' (1999) definition. A real-world example might be the management of the San Antonio Segment of the Edwards (Balcones Fault Zone) Aquifer where the Endangered Species Act (not the locals) resulted in the sustainable management of the aquifer. Under the Gleeson and others' (2020) definition, the sustainable management of the Edwards Aquifer would not be considered groundwater sustainability since inclusivity, equity, and long-term governance were not considered.

Perhaps because I am a hydrogeologist by training, I see sustainable production as being defined solely from the perspective of the aquifer—water is either being produced sustainably or not, regardless of the governance or lack thereof that arrived at that production. Alley and others' (1999) definition is silent on who exactly determines unacceptable environmental, economic, or social consequences. In my opinion, inclusivity, equity, and long-term governance should be recommended in the *process* to define unacceptable environmental, economic, or social consequences for assessments of groundwater sustainability, not in the definition of groundwater sustainability. There can be plenty of arguments over what exactly is considered inclusivity, equity, and long-term governance that are probably best left outside of the definition of groundwater sustainability.

Today Is the First Day of Forever

Since everybody and their third uncle has weighed in on safe yield (Table 6.1), sustainability, sustainable yield, and the superiority of cats and dogs, so shall I. Afterall, I read all (or nearly all) I could about safe yield and, at this moment (if just for a moment), probably know more

Table 6.1 List of terms for sustainable and unsustainable management of groundwater

- Terms for sustainable groundwater management
  - Ground Water Yield (Lee 1912)
  - Safe Yield (Lee 1915)
    - Yield (Charles 1946)
    - Perennial Yield (Williams and Lohman 1949)
      - Parker (1951)
      - Mann (1963)
    - Sustained Yield (McGuinness 1951)
      - Practical Sustained Yield (Suter and others 1959)
        - Walton (1964)
      - Mann (1963)
      - Maximum Sustained Yield (ASCE 1961)
      - Permissive Sustained Yield (ASCE 1961)
    - Optimum Yield (McGuinness 1951)
    - Basin Yield (Freeze 1971)
      - Natural Basin Yield (Freeze 1971)
      - Maximum Basin Yield (Freeze 1971)
        - Anderson and Berkebile (1977)
      - Basin Safe Yield (Freeze 1971)
      - Freeze and Cherry (1979)
    - Aquifer Yield (Freeze and Cherry 1979)
    - Reliable Yield (Khan and Mawdsley 1988)
    - Sustainable Yield (Downing 1993)
  - Ground-Water Sustainability (Alley and other 1999)
- Terms for nonsustainable groundwater management
  - Maximum Mining Yield (ASCE 1961)
    - Permissive Mining Yield (ASCE 1961)
    - Optimal Mining [Yield] (Renshaw 1963)
    - Optimal Mining Yield (Bear and Levin 1967)
- Terms that include sustainable and unsustainable management
  - Optimal Yield (Bear and Levin 1967)
    - Domenico (1972)
  - Groundwater Availability (TWDB 1968)

about it than anyone. I also bring my perspective as a practicing hydrogeologist working in groundwater management to the discussion.

In my opinion, much of the technical handwringing on safe yield has been much ado about nothing. As our understanding of aquifers and the technical sophistication of our tools have improved and as societal mores have become more enlightened over time (in part informed by our understanding and sophistication), safe yield's general defining feature of avoiding "dangerous" or "unacceptable" outcomes in the pursuit of sustainable pumping has proven amazingly resilient. Over time, the menu of unacceptable outcomes has expanded from maximizing production for human use to including environmental and societal impacts. Because of its long history in literature and practice, some may be using an older, and thus outdated, definition of safe yield. Safe yield should not bear the burden of the hydrologically unsophisticated incorrectly equating it to recharge or unsatisfactorily applying it to a management question. Safe yield doesn't kill aquifers; people kill aquifers.

Many in engineering and science have also wrung their hands over ambiguity in sustainable development and quantifying safe or sustainable yield; however, this ambiguity—as expressed through policy decisions—is an essential element of groundwater governance and management. People and processes are unpredictable, and that's OK. Science alone does not tell you how much to pump from your aquifer—science must work together with policy, almost always in a supporting role, answering questions, running what-if scenarios, and supporting stakeholders and policymakers in making decisions on what is acceptable and what is unacceptable. Once policymakers make policy decisions, science can then provide the best available answer on how much can be produced while avoiding the identified unacceptable outcomes.

In my opinion, Alley and others (1999) provide the best definition of groundwater sustainability as the "development and use of ground water in a manner that can be maintained for an indefinite time without causing unacceptable environmental, economic, or social consequences." A hairsplitting (but clever) groundwater district manager from the High Plains of Texas once argued with me after I revealed Alley and others' definition at a conference that his district was managing sustainably by that definition, since the development and use of the aquifer in his area would be maintained indefinitely albeit at an asymptotically lowering rate as the saturated thickness of the aquifer declined. Ultimately, once the economics of large-volume irrigation dried-up, production of the aquifer would subsequently be indefinite since users would only be able to produce what the aquifer was willing to give. This approach clearly has environmental, economic, and social consequences, but they are all acceptable to the majority his local users, users who elect the board members of his district who then set the production policy for the area.

I appreciate his logic, but his conclusions go against the intent of sustainability. It would also mean that all aquifers are managed sustainably (assuming that all unacceptable consequences—however they are defined—are being avoided). Alley and others (1999) use of the word "maintain" is intended to apply to the development and use of groundwater (if groundwater development and use decreases over time, is it really maintained?). Merriam-Webster's (2021) primary definition for maintain is "to keep in an existing state (as of repair, efficiency, or validity): preserve from failure or decline."

Groundwater sustainability is a policy preference informed by science. Imbedded in groundwater sustainability are policy decisions on what exactly are unacceptable environmental, economic, or social consequences. Some governing body has to make the policy decision (see Chap. 7 on groundwater governance). Sustainable yield is how much groundwater can be produced on an annual basis from an aquifer to achieve groundwater sustainability as constrained by the unacceptable environmental, economic, or social consequences. In Texas parlance, groundwater sustainability would be the desired future condition for an aquifer, while sustainable yield is how much groundwater can be produced from that aquifer to achieve that desired future condition (the modeled available groundwater).

Sustainable yield is safe yield by a different name. Despite the misunderstandings of safe yield over the years, my preference is to use sustainable yield because, in general, hydrogeologists all know what this means. Many do not understand what safe yield means and how it has changed over time. There is also something to be said about abandoning a term that is also used in surface water but means something different. Using sustainable yield also avoids the use of the word "safe," which, as others have noted, has ambiguous meaning and is not necessarily safe. If someone uses safe yield, you will want to see what definition they are using and why. In some cases, safe yield is baked into law. And if someone is using the modern (dare I say, enlightened?) definition of safe yield, there is no need to wag your finger in their face.

Sustainable production is a general term that can be used to represent any level of groundwater production that can be performed indefinitely. Sustainable yield is a special case of sustainable production. Maximum sustainable production is the maximum amount of groundwater that can be indefinitely produced independent of any undesired affects. Therefore, sustainable yield can be between zero and the maximum sustainable production.

In summary:

Groundwater sustainability is the "development and use of ground water in a manner that can be maintained for an indefinite time without causing unacceptable environmental, economic, or social consequences" (Alley and others 1999).

Sustainable yield is how much groundwater can be produced on an annual basis from an aquifer to achieve groundwater sustainability as constrained by unacceptable environmental, economic, or social consequences.

Sustainable production refers to any amount of groundwater production that can be achieved indefinitely.

Safe yield is the same as sustainable yield.[4]

Questions

What would you consider to be unacceptable environmental, economic, or social consequences? Why?

Are you surprised at how long it took for scientists and engineers to consider impacts to the environmental as part of safe yield? Why?

What is your preferred term for safe yield? Why?

How long is forever?

References

Alley, W.M., and Leake, S.A., 2004, The journey from safe yield to sustainability: Ground Water, v 49, n 1, p 12–16.

Alley, W.M., Reilly, T.E., and Franke, O.E., 1999, Sustainability of groundwater resources: U.S. Geological Survey Circular 1186, Denver, Colorado, 79 p. https://pubs.usgs.gov/circ/circ1186/.

---

[4] Almost forgot: cats are superior to dogs.

Anderson, M.P., and Berkebile, C.A., 1977, Hydrogeology of the South Fork of Long Island, New York – Discussion: Geological Society of America Bulletin, v. 88, p. 895.

ASCE (American Society of Civil Engineering), 1961, Ground water basin management: Manual of Engineering Practice No. 40, 160 p.

ASCE (American Society for Civil Engineers), 1977, Policy Statement 243—Groundwater Management: American Society for Civil Engineers https://www.asce.org/issues-and-advocacy/public-policy/policy-statement-243%2D%2D-groundwater-management/.

ASCE Task Committee on Sustainability Criteria. 1998. Sustainability Criteria for Water Resource Systems. American Society for Civil Engineers, Reston, Virginia, USA. 253 pages.

Bear, J., and Levin, O., 1967, The optimal yield of an aquifer: International Association of Scientific Hydrology Bulletin, Publication 72, p. 401–412.

Bredehoeft, J.D., 1997, Safe yield and the water budget myth: Ground Water, v. 35, n. 6, p. 929.

Bredehoeft, J.D. 2002. The water budget myth revisited: Why hydrogeologists model. Ground Water, v. 40, n. 4, p. 340–345.

Bredehoeft, J.D., Papadopulos, S.S., and Cooper, H.H., 1982, Groundwater – The water budget myth: *in* Scientific Basis of Water-Resource Management, Studies in Geophysics, Washington, DC: National Academy Press, p. 51–57.

Brinkmann, R., 2020, *Environmental Sustainability in a Time of Change*: Palgrave Studies in Environmental Sustainability, Palgrave Macmillan, 325 p.

Brown, L.R., 1982, Building a sustainable society: Society, v. 19, p 75–85.

CG-ASCE (Committee on Groundwater for the American Society of Civil Engineers), 1961, Ground water basin management: Manual of Engineering Practice No. 40, 160 p.

Charles, J.R., 1946, Discussion on "Utilization of groundwater storage in stream system development": Transactions of the American Society of Civil Engineers, v 111, n 1, p 319–326.

Conkling, H., 1946, Utilization of groundwater storage in stream system development: Transactions of the American Society of Civil Engineers, v 111, n 1, p 33–62.

Devlin, J.F., Sophocleous, M., 2004, The persistence of the water budget myth and its relationship to sustainability. Hydrogeology Journal, v. 13, 549–554.

Domenico, P.A., 1972, Concepts and Models in Groundwater Hydrology: McGraw-Hill, New York, 405 p.

Domenico, P.A., and Schwartz, F.W., 1990, Physical and Chemical Hydrogeology: John Wiley & Sons, 824 p.

Downing, R.A. 1993. Groundwater resources, their development and management in the UK: An historical perspective: Quarterly Journal of Engineering Geology, v. 26, p. 335–358.

Evans, W.R., and Cook, P.G., 2002, What is a sustainable yield for Australia's groundwater systems? International Association of Hydrogeologists Darwin Conference: Balancing the Groundwater Budget, Darwin, Australia, May 12–17.

Fetter, Jr., C.W., 1972, The concept of safe groundwater yield in coastal aquifers: Water Resources Bulletin, v. 8, n. 6, p. 1173–1176.

Fetter, Jr., C.W., 1976, Hydrogeology of the South Fork of Long Island, New York: Geological Society of America Bulletin, v. 87, p. 401–406.

Fetter, Jr., C.W., 1977, Hydrogeology of the South Fork of Long Island, New York – Reply: Geological Society of America Bulletin, v. 88, p. 896.

Fetter, Jr., C.W., 1980, Applied Hydrogeology (1st Edition): Prentice-Hall, 544 p.

Freeze, R.A., 1971, Three-dimensional, transient, saturated-unsaturated flow in a groundwater basin: Water Resources Research, v. 7, n. 2, p. 347–366.

Freeze, R.A., and Cherry, J.A., 1979, *Groundwater*: Prentice-Hall, Inc., 604 p.

Gleeson, T., Alley, W.M., Allen, D.M., Sophocleous, M.A., Zhou, Y., Taniguchi, M., and VanderSteen, J., 2012, Towards sustainable groundwater use: Setting long-term goals, backcasting, and managing adaptively: Groundwater, v. 50, n. 1, p. 19–26.

Gleeson, T., Cuthbert, M., Ferguson, G., and Perrone, D. 2020, Global groundwater sustainability, resources, and systems in the Anthropocene: Annual Review of Earth and Planetary Science, v. 48, p. 431–463.

Gleeson, T., VanderSteen, J., Sophocleous, M.A., Taniguchi, M., Alley, W.M., Allen, D.M., and Zhou, Y., 2010, Groundwater sustainability strategies. Nature Geoscience v. 3, n. 6, p. 378–379.

Gleeson, T., Villholth, K., Taylor, R., Perrone, D., Famiglietti, J., Hyndman, D., van Loon, A., and Wada, Y., (leadership team), 2019, Global groundwater sustainability – A call to action: Global Groundwater Statement, groundwaterstatement.org, 2 p.

Hamlin, S.N., and Anthony, S.S., 1987, Ground-water resources of the Laura Area, Majuro Atoll, Marshall Islands: U.S. Geological Survey, Water Resources Investigations Report 87-4047, 69 p.

Hiscock, K.M., M.O. Rivett, and R.M. Davison, 2002. Sustainable groundwater development: in Sustainable Groundwater Development: Special Publication No. 193. London: Geological Society, p 1–14.

Hopkins, D., 1987, Management of the Gnangara Mound groundwater resources [Western Australia]: International Commission on Irrigation and Drainage. Australian National Committee, Australian National Committee, p. 1.24–1.34.

IUCN (International Union for Conservation of Nature and Natural Resources), 1980, *World Conservation Strategy – Living Resource Conservation for Sustainable Development*: International Union for Conservation of Nature and Natural Resources, United Nations Environment Programme, and the World Wildlife Fund; 77 p.

Kalf, F.R.P., Woolley, D.R., 2005. Applicability and methodology of determining sustainable yield in groundwater systems. Hydrogeology Journal 13, 295–312.

Kazmann, R.G., 1956, "Safe yield" in ground-water development, reality or illusion? Journal of the Irrigation and Drainage Division, Proceedings of the American Society of Civil Engineers, Paper 1103, 12 p.

Khan, L.R., and Mawdsley, J.A., 1988, Reliable yield of unconfined aquifers: Hydrological Sciences Journal, v33, no. 2, p. 151–171.

Lee, C.H., 1912, An intensive study of the water resources of a part of Owens Valley, California: U.S. Geological Survey Water-Supply Paper 294, 144 p.

Lee, C.H., 1915, The determination of safe yield of underground reservoirs of the closed basin type: Transactions of the American Society of Civil Engineers, v 78, n 1, p 148–151.

Loaiciga, H.A., 2016, The safe yield and climatic variability—Implications for groundwater management: Groundwater, v 55 n 3, p. 334–345.

-Mace, R.E., 2021, Five gallons in a ten gallon hat – Groundwater sustainability in Texas: The Meadows Center for Water and the Environment, Texas State University, Report 2021-08, 50 p.

Maimone, M., 2004, Defining and managing sustainable yield: Ground Water, v 42, n 6, p 809–814.

Mann, J.F., Jr., 1961, Factors affecting the safe yield of ground-water basins: Journal of the Irrigation and Drainage Division, Proceedings of the American Society of Civil Engineers, v 87, n 3, p 63–69.

Mann, J.F., Jr., 1963, Factors affecting the safe yield of ground-water basins: Transactions of the American Society of Civil Engineers, v 128, Paper No. 3434, p 180–186.

Mann, J.T., Jr., 1968, Concepts in ground water management: Journal of the American Water Works Association, v 60, n 12, p 1336–1344.

McGuinness, C.L., 1951, The water situation in the United States with special reference to ground water: U.S. Geological Survey Circular No. 114, 127 p.

Meinzer, O.E., 1920, Quantitative methods of estimating ground-water supplies: Bulletin of the Geological Society of America, v 31, p 329–338.

Merriam-Webster, 2021, Maintain: in Merriam-Webster.com dictionary. Retrieved September 30, 2021, from https://www.merriam-webster.com/dictionary/maintain.

NGC (National Groundwater Committee), 2004, Knowledge gaps for groundwater reform: in Proceedings from Workshop 12–13, November 2003, Canberra, Australia.

Parker, G.G., 1951, Geologic and hydrologic factors in the perennial yield of the Biscayne Aquifer: Journal of the American Water Works Association, v 43, p 817–834.

Peralta, R., Timani, B., and Das, R., 2011, Optimizing safe yield policy implementation: Water Resources Management, v 25, p 483–508.

Pierce, S.A., Sharp, Jr., J.M., Guillaume, J.H.A., Mace, R.E., and Eaton, D.J., 2012, Aquifer-yield continuum as a guide and typology for science-based groundwater management: Hydrogeology Journal, v 21, p 331–340.

Price, M., 2002, Who needs sustainability? *in* Hiscock, K.M., Rivett, M.O., and Davison, R.M., editors, *Sustainable Groundwater Development*: Geological Society, London, Special Publications, 193, p 75–82.

Renshaw, E.F., 1963, The management of ground water reservoirs, Journal of Farm Economics, v 45, 285–295.

Rudestam, K., and Langridge, R., 2014, Sustainable yield in theory and practice – Bridging scientific and mainstream vernacular: Groundwater, v. 52, p. 90–99.

Sophocleous, M., 1997, Concluding comments on managing water-resources systems – Why "safe yield" is not sustainable in Perspectives on Sustainable Development of Water Resources in Kansas, Kansas Geological Survey Bulletin 239, Chapter 10, p. 211–213.

Sophocleous, M., 2000, From safe yield to sustainable development of water resources – The Kansas experience. Journal of Hydrology, v 235, n 1–2, p 27–43.

Suter, M., Bergstrom, R.E., Smith, H.F., Emrich, G.H., Walton, W.C., and Larson, T.E., 1959, Preliminary report on ground-water resources of the Chicago Region, Illinois: Illinois State Water Survey and Geological Survey Cooperative, Ground-Water Report 1, 88 p.

Theis, C.V., 1940, The source of water derived from wells—Essential factors controlling the response of an aquifer to development: Civil Engineer, v 10, p 277–280.

Thomas, H.E., 1951, The Conservation of Groundwater: McGraw-Hill Book Company, New York.

Thomas, R.O., 1959, Legal aspects of ground water utilization: Proceedings of the American Society of Civil Engineers, v 85, n 4, p 41–63.

Todd, D.K., and Meyer, C.F, 1970, A study of the hydrology and geology of Area I of the Honolulu Aquifer: prepared for Board of Water Supply City and County of Honolulu, Completion Report 70-TMP-20, 79 p.

Todd, D.K., and Meyer, C.F, 1971, Hydrology and geology of the Honolulu Aquifer: Proceedings of the American Society of Civil Engineers, Journal of the Hydraulics Division, v. 97. No. 2, p. 233–256.

Tolman, C.F., 1937, Ground Water: McGraw-Hill Book Co., Inc., New York, N. Y., 1937, 593 p.

Tóth, J., 1973, Hydrogeology and yield evaluation of a municipal well field, Alberta, Canada: Hydrological Sciences Journal, v. 18, no. 2, p. 165–189.

TWDB (Texas Water Development Board), 1968, The Texas water plan: Texas Water Development Board, Austin, Texas; variously paginated (252 total pges).

United Nations, 1992, Agenda 21: United Nations Conference on Environment & Development, Rio de Janerio, Brazil, 3 to 14 June 1992, 351 p.

Walton, W.C., 1964, Future water-level declines in deep sandstone wells in Chicago region: Ground Water, v 2, n 1, p 13–20.

WCED (World Commission on Environment and Development), 1987, *Our Common Future*: Oxford University Press, 400 p.

Williams, C.C., and Lohman, S.W., 1949, Geology and ground-water resources of a part of south central Kansas: Geological Survey of Kansas, Bulletin 79, 212 p. http://www.kgs.ku.edu/Publications/Bulletins/79/index.html.

Wood, W.W., 2001, Water sustainability – Science or management? Ground Water, v 39, n 5, p 641.

Worster, D., 1993, The shaky ground of sustainability *in* Sachs, W., editor, Global Ecology – A New Era of Political Conflict: Zed Books, p 133–145.

Zhou, Y., 2009, A critical review of groundwater budget myth, safe yield, and sustainability: Journal of Hydrology, v 370, p. 207–213.

7

Groundwater Governance and Sustainability

"Everyone needs a seat at the water table."
International Union for Conservation of Nature (Smith and others *2016*)

For groundwater sustainability, science informs, but policy decides. Science provides the best understanding of the time of an aquifer and how that aquifer responds to production and use. One controlling factor that science provides in groundwater sustainability discussions are the boundaries of the sustainability playing field. One boundary is easy to define: zero production, a level of production that maintains water levels and spring and base flows in the aquifer in its original state or attempts to return an aquifer to its original state. Technically, sustainable production could be negative; in other words, putting more water into the aquifer than nature provides such as through artificial recharge projects, but this is uncommon (or unheard of—I am not aware of a single case). The other boundary is the maximum sustainable production independent of environmental, economic, or social consequences.

Outside of setting the scientific playing field, policy is the controlling factor for groundwater sustainability. Policy decisions define the unacceptable environmental, economic, or social consequences. Science

R. E. Mace, *Groundwater Sustainability*, Palgrave Studies in Environmental Sustainability, https://doi.org/10.1007/978-3-031-13516-3_7

defines the limits of what is physically possible and estimates the outcomes of policy preferences, but it is people that ultimately decide whether or not aquifers are managed sustainably and, if managed sustainably, at what level of unacceptable consequences.

Policy is a set of ideas or a plan of what to do in particular situations, which has been agreed to officially by a group of people or a government (modified from Cambridge Dictionary 2021). *Groundwater governance* is "the decision of which management actions should be taken, when, by who, and for what purpose" while *groundwater management* is the implementation of that decision (Villholth and Conti 2019). Groundwater governance is the process of decision-making, policy is the outcome of the governance process, and management is the implementation of the policy outcome. The purpose of this chapter is to describe how this policy pathway of process, outcomes, and implementation works and how it relates to groundwater sustainability.

Groundwater Governance

Groundwater governance is a relatively recent term introduced in the late 2000s (Foster and others 2009) building on concepts of general governance that started in the 1980s and developed over time (Villholth and Conti 2019). Foster and others (2009) defined groundwater governance as "the exercise of appropriate authority and promotion of responsible collective action to ensure sustainable and efficient utilization of groundwater resources for the benefit of humankind and dependent ecosystems." Later definitions, including that by Megdal and others (2015), removed the normative elements from this and other earlier definitions. Being normative means imparting a value judgment. For example, the Foster and others (2009) definition is value-laden through the use of the words "appropriate," "responsible," "sustainable," and "efficient." That is not to say that these are not reasonable outcomes for most people, but groundwater is still governed even if, for example, it is not managed sustainably. There is also plenty of discussion about what "good" groundwater governance is, another term that is normative and is not used anymore by enlightened experts.

Global Environment Facility (2015) presented four components of groundwater governance: (1) actors, (2) legal framework, (3) policies, and (4) information and knowledge.

Actors

Actors include anyone involved in (1) using (or potentially using) groundwater, (2) being affected by the use of groundwater, (3) legalities, (4) policies and planning, and (5) science. The stakeholders—those with an interest or concern in groundwater—are generally the existing users, potential users, or others affected by groundwater use and policy outcomes. Stakeholders also include actors affected by the use and potential use of groundwater such as surface-water users; organizations involved in environmental, economic, social issues; and landowners potentially affected by groundwater use and policy.

Actors involved in legalities may include legislative and congressional bodies, the courts, attorneys that represent various stakeholders, and attorneys representing governmental bodies that pass and interpret rules. Actors involved in policy and planning may include legislative and congressional bodies, board members, and other elected officials as well as bureaucrats charged with implementing policy and planning and the associated processes for developing policies and plans. Actors involved in science include nearly all the aforementioned but also include scientists which may be academic, governmental, or private.

While science should be conducted independent of policy preferences, governmental bodies determine budgets for the acquisition of science, what science is conducted, who conducts the science, and how the science is developed and shared with stakeholders and the public. Stakeholders determine how they each use and interpret the science. Science and the obfuscation of science is often used as weapons during intense groundwater policy battles.

Legal Frameworks

Legal frameworks generally define ownership, regulation of groundwater production and use, and responsibilities which set the boundaries for what is legally possible for groundwater management. Legal frameworks are often established by legislative, congressional, or parliamentary bodies through constitutions and statutes or by the courts. Absent direct legislative or congressional action, courts may default to common law. For example, in Texas, the state supreme court, when faced with a groundwater production interference case in 1904 and finding nothing in statute to decide the case, defaulted to the English Common Law's Rule of Capture. Legal frameworks may also be set by the interpretation of holy books in theocracies or by fiat in dictatorships.

Early constitutional, legislative, and legal decisions can greatly influence today's groundwater governance. For example, the Texas Supreme Court's decision in 1904 still provides the legal foundation for groundwater law in the state. Early decisions on water memorialized in constitutions may also continue to be relevant today. As human constructs, legal frameworks can be changed, but the changeability of these frameworks is often defined by politics, which can be unpredictable, and policy inertia, which can be difficult to redirect.

In most parts of the world, groundwater has been declared the property of the state and regulated as such although there can be a protected legacy of previous ownership rights (Global Environment Facility 2015). There are numerous groundwater doctrines in use. In the United States, common doctrines include the Rule of Capture, Reasonable Use, Correlative Rights, Prior Appropriation, and Second Restatement of Torts.

Rule of Capture

The Rule of Capture is the oldest of the doctrines with a history—at least for wild animals—that reaches back to Greek times (Wise 1996; Drummond and others 2006). The Rule of Capture was first invoked for groundwater in Roman times when the Justinian Digest summarized

Marcus Claudius Marcellus (?-45 BC) as concluding that "no action, not even the action for fraud, can be brought against a person who, while digging on his own land, diverts his neighbor's water supply" (Watson and others 1985; Drummond and others 2004). The Rule of Capture, whether explicitly adopted or not, has been the customary law for groundwater in many parts of the world and is still significant for small-scale production in rural areas of developing countries (Global Environment Facility 2015) and parts of the United States.

In modern times, England's Acton v. Blundell (1843) is referred to as the source for English Common Law. In this case, a mining company allegedly drained a water well through the action of a nearby mining operation. The court concluded that "the existence and state of underground water is generally unknown before a well is made; and after it is made there is the difficulty of knowing exactly how much, if any, of the water of the well, when the ground was in its natural state, belonged to the owner in right of his property in the soil, and how much belonged to his neighbor. These practical uncertainties made it very reasonable not to apply the rules which regulate the enjoyment of streams and waters above ground to subterranean waters" and, as a consequence, ruled that "the person who owns the surface may dig therein, and apply all that is there found to his own purposes at his free will and pleasure; and that if, in the exercise of such right, he intercepts or drains off the water collected from underground springs in his neighbour's well, this inconvenience to his neighbour falls within the description of *damnum absque injuriâ* [damage without injury], which cannot become the ground of an action."

Also referred to as the Absolute Ownership Doctrine, the Absolute Dominion Rule, and the Common Law Rule (and informally as the Law of the Biggest Pump), the Rule of Capture generally states that the person who owns the surface also owns the groundwater beneath the surface and can withdraw that groundwater at any rate regardless of harm to neighbors, streams, and springs. The Rule of Capture operates in the de facto sense before any doctrine is adopted because the outcome is the same absent any regulation: each landowner gets to do what they want on their property. Once the Rule of Capture is formally chosen as the governing doctrine, landowners can still do whatever they want—absent malice and waste—but now with case law supporting their actions.

Needless to say, it is nearly impossible to produce groundwater sustainably under an unmodified Rule of Capture unless the production happens to be less than the maximum sustainable production and is expected to remain so indefinitely into the future. It's important to note that the Rule of Capture operates at several levels: (1) tort law, (2) ownership, and (3) rules of production. Tort law relates to the ability to sue (or, more accurately, the ability to sue and win damages), ownership relates to who owns the water (the landowner), and rules of production relates to how much groundwater is produced from where. Texas, widely recognized as a Rule of Capture state, has allowed for the creation of groundwater conservation districts, which supersede the Rule of Capture with respect to the rules of production. Under the Rule of Capture, there are no rules of production (barring producing purely out of malice). In Texas, groundwater conservation districts can control well spacing from property lines and from other wells and can limit how much is produced, thereby superseding the Rule of Capture with respect to the rules of production (Mace 2019).

Reasonable Use

In the late 1840s, the Salisbury Manufacturing Company built a dam on the Powow River in New Hampshire (Dellapenna 2013). As the reservoir behind the dam filled, it raised the water table in the bordering aquifer, waterlogging property owned by a Dr. Bassett which prevented him from farming. Interestingly, this was not a case of one landowner lowering water levels in an aquifer beneath another landowner and causing damage; instead, it was one landowner *raising* the water levels beneath another landowner and causing damage.

Dr. Bassett filed suit, prompting a 15-year legal battle with five trials that ultimately resulted in the New Hampshire Supreme Court abolishing the Rule of Capture in the state and replacing it with what became known as the Reasonable Use Doctrine (Dellapenna 2013; Collins 2021). The Court concluded that "No land-owner has an absolute and unqualified right to the unaltered natural drainage or percolation to or from his neighbor's land" and that "[t]he rights of each land-owner being similar, and his enjoyment dependent upon the action of the other land-owners, these rights must be valueless unless exercised with reference to each other" (Bartlett 1862).

Also known as the American Rule and the Qualified Ownership Doctrine, this legal framework generally holds that (1) the landowner owns the water beneath their property, (2) the landowner cannot move groundwater off of their property (by definition, this is generally considered unreasonable), and (3) groundwater discharge to surface-water bodies is recognized and protected. For example, Illinois defines reasonable use for both groundwater and surface water as "the rule of water to meet natural wants and a fair share for artificial wants" (Janasie 2020).

Although landowners own groundwater beneath their property under the Reasonable Use Doctrine, they do not have the right to produce as much as they want nor can they unreasonably effect their neighbors or other natural resources, so their ownership is qualified. The key policy question is: What is reasonable? If reasonableness includes impacts to rivers, streams, and springs, then the Reasonable Use Doctrine can result in sustainable production from aquifers.

Correlative Rights

After Red Lake River became contaminated, the City of Crookston in Minnesota drilled wells into the artesian aquifer beneath its streets (Wenzell 1909). Consequently, the city's pumping reduced water levels in the aquifer, causing a flowing artesian well owned by Mr. Erickson to stop flowing, forcing him to install and operate a pump (Wenzell 1909). Mr. Erickson sued the city for $3000 in damages due to their pumping. In its ruling, the Minnesota Supreme Court established correlative rights to groundwater (Wenzell 1909).

Also called the Fair Use Doctrine (and sometimes referred to as Reasonable Use), the Correlative Rights Doctrine correlates groundwater rights to, generally, land surface area above an aquifer. In other words, how much you can produce from an aquifer depends on how much land you own over an aquifer. Sometimes hydrogeologic properties are incorporated into correlative rights. For example, a groundwater conservation district in Texas correlates its groundwater rights to both surface acreage and sand thickness (Blumberg and Collins 2016).

Under Correlative Rights, the landowner owns the groundwater, but the right is not absolute and is, instead, relative to other users who are coequal such that everyone gets their fair share. Correlative Rights are generally allocated by the courts but can also be done legislatively and through local regulators (such as described in Blumberg and Collins 2016). Groundwater under correlative rights can be transported off of the property.

Depending on overall management goals, correlative rights may or may not result in sustainable management of groundwater. Correlative rights to achieve sustainability can be challenging because the amount of groundwater that can be produced, once divided across the broad extent of an aquifer, is generally small compared to unfettered, unsustainable production.

Prior Appropriation

After James W. Marshall discovered gold at Sutter's Mill in Coloma, California, on the Monday morning of January 24th, 1848 (Rohrbough 2013), more than 150,000 people rushed to the Golden State over the next two years seeking fame and fortune. Panning for gold required water, and lots of it. By 1882, there were over 6000 miles of ditches and canals—including the 247-mile-long Eureka Canal—constructed to assist in gold exploration (Miller 1973). This unprecedented tapping of surface waters combined with the lack of state or federal regulation resulted in the miners adopting their own water management approach: "First come, first served," which turned into "First in time, first in right," which turned into the Prior Appropriation Doctrine (Miller 1973).

Under prior appropriation, groundwater is generally owned by the state, but the permits are private property, which can be bought and sold. Prior appropriation for groundwater generally occurs in states where surface water is also managed through prior appropriation. In cases where the state recognizes the connection between groundwater and surface water, groundwater is often junior to surface water because historically people used and states permitted surface water first. Colorado recognizes tributary and non-tributary groundwater, where tributary groundwater contributes flows to surface water (and are thus "tributary" to rivers and streams) and non-tributary groundwater does not. Under prior appropriation, the amount of groundwater permitted may be limited to a sustainable yield, but it is not required.

Second Restatement of Torts

In 1923, a group of attorneys, judges, and legal scholars formed the American Law Institute to clarify, modernize, and improve United States common law. In 1965, the Institute published the *Restatement of the Law Second, Torts* which included a common law hybrid of the Rule of Capture and the Reasonable Use Doctrine for groundwater. Second Restatement of Torts (for groundwater) retains the ownership and non-liability aspects from the Rule of Capture and the reasonable use aspect from the Reasonable Use Doctrine. Since reasonable use generally provides for the protection of baseflows to rivers and streams, Second Restatement of Torts may accommodate sustainability management goals for groundwater.

Policies

Policy is a principle or course of action adopted as desirable, advantageous, or expedient by a government (after Oxford University Press 2021). In groundwater governance, policies are "decisions made by stakeholders regarding what to do in the context of the governance framework" (Global Environment Facility 2015), although policy is also important in establishing or changing the governance framework. In the traditional definition, stakeholders (generally regulatory bodies that interact with scientists, users, and other interests) work within the legal framework to identify management goals and the specific steps to achieve those goals.

In countries with established legal doctrines, policies have to exist within those legal constraints; however, policies can also develop to change those doctrines (e.g., moving from Rule of Capture to Second Restatement of Torts [Collins 2021]). In countries with no existing legal doctrine, then developing a national groundwater management vision (Smith and others 2016) can inform policies and the development of an appropriate legal doctrine.

Information and Knowledge

At first blush, it may not be clear what information and knowledge—facts or what we understand the facts to be—have to do with groundwater governance. However, people and policy often dictate what information is collected, how much of it is collected, and how it is shared, and information and knowledge are often critical in making policy, governance, and management decisions. And you can't manage what you don't measure.

Groundwater decisionmakers and managers can employ a wide range of information on their aquifers to make decisions, including spring and well surveys, regular water-level and water-quality monitoring, geologic mapping at the surface and in the subsurface, geochemical studies, gain-loss studies on rivers and streams, permeability testing, and the development of numerical groundwater flow models (Mace 2018). Decisionmakers determine which studies and monitoring should be done and at what level. These decisions may be restricted by budgetary limitations, which may source back to stakeholders who pay fees and taxes to support scientific work.

Decisionmakers also decide at what level stakeholders participate in the prioritization and collection of data and how the resulting information is shared. Does the collected information follow open data protocols and shared with the public in a timely manner (Charalabidis and others 2018)? Does the regulatory body strive to clearly communicate the science in an understandable way to the regulated body? Even the development of numerical groundwater flow models can have stakeholder involvement to increase the effectiveness of management tools (Mace and

Ridgeway 2016). Regulatory bodies can increase trust by making the information they use available for public scrutiny. Furthermore, third parties, such as academic researchers and the private sector, can leverage available information to better understand the aquifer at no cost to the regulators.

Information and knowledge do not only relate to science. Permits and processes are also information and knowledge as are meetings to make decisions. Regulatory bodies can strive to make this information and knowledge easily accessible to the regulated body and other interested parties.

Lack of awareness about groundwater (out of sight, out of mind) and its importance also affects overall groundwater governance (Global Environment Facility 2015). If people are not concerned about groundwater, it can be difficult to get them to agree to be regulated and pay for that regulation and the development of information and knowledge to facilitate that regulation.

Groundwater Management

As Villholth and Conti (2019) noted, groundwater management is the implementation of the decision of which management actions should be taken, when, by who, and for what purpose. Here, too, governance exerts control since governance also controls the amount of funding available to facilitate implementation. For example, an understaffed regulatory body may not be able to implement the policies and regulations dictated by the governance structure.

Groundwater management often requires a general manager to lead the implementation of identified management actions and, if the governance structure extends to the management level, interact with the decisionmakers on creating and modifying policies and decisions. Other staff may include technicians to collect data, scientists to interpret data to advance knowledge of the resource, specialists to process permits, attorneys to ensure that all actions are in compliance with the rules and regulations of the governance structure, communication specialists to share information with the press and public, and associated support staff.

Various consultants and experts may also be employed to assist the general manager in achieving the goals and requirements set by the governance structure.

Open information improves transparency but also allows stakeholders to police management actions. Is the groundwater management achieving what the governance expects it to achieve? Is the managing authority achieving their data quantity and quality objectives? How is the regulated community reporting their data (and how is it validated)?

Groundwater Governance for Sustainability

Groundwater governance does not, by definition, necessarily result in sustainable management of aquifers. For example, the Ogallala Aquifer has a well-developed groundwater governance and management structure, but it is not, as a whole, sustainably managed. But just as groundwater governance is defined by its actors, legal framework, policies, and information and knowledge, so is sustainable groundwater governance.

Sustainable groundwater governance requires actors that want and actively seek the sustainable management of aquifers (or, if it happened in the past, wanted and actively sought such management). Depending on the overall groundwater governance structure, these actors may be needed at the federal, state, regional, or local level or at a combination of levels. In other words, somebody (preferrable somebody important and/or the general public) has to want to manage their aquifers sustainably.

Sustainable groundwater governance requires a legal framework that supports sustainable management of aquifers. Some legal doctrines directly support sustainable management, some support sustainable management as an un intended side effect of protecting groundwater contributions to surface water, and some don't support it at all. In the latter case, there then must be actors seeking to change the legal doctrine into something that requires or at least allows sustainable management of groundwater.

Sustainable groundwater governance requires policies that support sustainable management of aquifers. In some cases, the legal doctrine may require sustainable management but state, regional, and/or local actors have to flesh out the policies of what sustainable means, how to measure it, and how to get there. In other cases where the legal doctrine allows but doesn't require sustainable management, then motivated actors need to convince the appropriate stakeholders and constituents that sustainable groundwater management is desirable.

Sustainable groundwater governance requires information and knowledge. Actors need to hold stakeholder meetings to identify sustainability goals and support the development of science that allows experts to assess sustainable yields based on those goals.

The next chapter provides several examples of governance and management to achieve sustainable groundwater management.

Questions

Why do academics consider governance and management separately?

What is the groundwater governance structure in your locale? Who are the actors?

Does your groundwater governance structure and actors support sustainable groundwater management?

Why or why not?

Do you have enough information and knowledge to manage your groundwater resources sustainably?

References

Bartlett, J., 1862, Basset v. Salisbury Manufacturing Company – Decision: New Hampshire Supreme Court, 43 New Hampshire 328.

Blumberg, H., and Collins, G., 2016, Implementing three-dimensional groundwater management in a Texas groundwater conservation district: Texas Water Journal, v. 7, n. 1, p. 69–81.

Cambridge Dictionary, 2021, Policy: https://dictionary.cambridge.org/dictionary/english/policy accessed October 1, 2021.

Charalabidis, Y., Zuiderwijk, A., Alexopoulos, C., Janssen, M., Lampoltshammer, T., and Ferro, E., 2018, The World of Open Data – Concepts, Methods, Tools and Experiences: Springer Nature, 229 p.

Collins, G., 2021, Overruling the Rule of Capture – What can Texas learn from 10 other states' groundwater law updates: Center for Energy Studies at Rice University's Baker Institute for Public Policy and The Meadows Center for Water and the Environment at Texas State University, 56 p., https://doi.org/10.25613/knsb-6g39.

Dellapenna, J.W., 2013, A primer on groundwater law: Idaho Law Review, v. 49, p. 265–323.

Drummond, D.O., Sherman, L.R., and McCarthy, Jr., E.R., 2004, The Rule of Capture – Still so misunderstood after all these years: Texas Tech Law Review, v. 37 p. 1–97.

Foster, S., Garduño, H., Tuinhof, A., and Tovey, C., 2009, Groundwater Governance: Strategic Overview Series No. 1. Washington, D.C.: World Bank.

Global Environment Facility, 2015, Global Diagnostic on Groundwater Governance: Special Edition for World Water Forum 7, 186 p. https://groundwaterportal.net/resource/global-diagnostic.

Janasie, C., 2020, An overview of water law in Illinois: Sea Grant Law Center, February 2020, NSGLC-20-04-02, 11 p.

Mace, R.E., 2018, Groundwater Management: in Groundwater – State of the Science and Practice, National Ground Water Association Press, Westerville, Ohio, p. 38–40.

Mace, R.E., 2019, Groundwater governance, management, regulation, and emerging issues in the Lone Star State: Water Resources IMPACT, American Water Resources Association, p.11–12.

Mace, R.E., and Ridgeway, C.K., 2016, Forecasting underground rain – Groundwater availability modeling: in Sahs, M.K., editor, Essentials of Texas Water Resources, TexasBarBooks, State Bar of Texas, Chapter 19, p. 423–438.

Megdal, S.B., Gerlak, A.K., Varady, R.G., and Huang, L.Y., 2015, Groundwater governance in the United States: Common priorities and challenges: Groundwater, v 53, n 5, p 677–684. https://doi.org/10.1111/gwat.12294.

Miller, G.R., 1973, Shaping California water law – 1781 to 1928: Southern California Quarterly, v. 55, n. 1, p. 9–42.

Oxford University Press, 2021, Policy: Oxford English Dictionary, accessed November 30, 2021, https://www.oed.com/view/Entry/146842?result=1&rskey=YjloBU&.

Rohrbough, M.J., 2013, Rush to Gold – The French and the California Gold Rush, 1848–1854: Yale University Press, 352 p.

Smith, M., Cross, K., Paden, M. and Laban, P., editors, 2016, Spring – Managing groundwater sustainably: International Union for Conservation of Nature, Gland, Switzerland, 133 p.

Villholth, K.G., and Conti, K.I., 2019, Groundwater governance – Rationale, definition, current state and heuristic framework: *in* Villholth, K.G., López-Gunn, E., Conti, K.I., Garrido, A., and van der Gun, J., Advances in Groundwater Governance: Taylor and Francis Group, p 3–31.

Watson, A., Mommsen, T., and Krueger, P., 1985, The Digest of Justinian: University of Pennsylvania Press; Philadelphia, Pennsylvania.

Wenzell, H.B., 1909, Ed. Erockson v. Crookston Waterworks, Power, & Light Company: Minnesota Reports—Cases Argued and Determined in the Supreme Court of Minnesota, v. 105, p. 182–198.

Wise, S.M., 1996, The legal thinghood of nonhuman animals: Boston College of Environmental Affairs Law Review, v 23, n 3, p. 471–547.

8

Pathways to Groundwater Sustainability

"Groundwater is always an issue of mine, ours, or yours. If I'm pumping it, it's mine. If you're pumping it, it's ours.If it's contaminated, it's yours."
Gregory Ellis, Texas Water Attorney, describing perceptions of groundwater ownership

There is not a single pathway to achieving groundwater sustainability, but some pathways are easier to follow than others. The easiest path to follow is if the legal framework already requires some semblance of the sustainable production of groundwater. For example, some states in the United States adopted the concept of safe yield in the early 1900s based on the recommendations of engineers and scientists. Other legal doctrines, through the protection of groundwater flows to the surface whether intentionally (Reasonable Use Doctrine) or not (Prior Appropriation), may also result in sustainable groundwater production.

Legal doctrines that are not supportive of sustainable production may have been amended through constitutional amendments and legislation to allow sustainable management. With legal doctrines that do not clearly support sustainability, policies can be critical in supporting sustainable

R. E. Mace, *Groundwater Sustainability*, Palgrave Studies in Environmental Sustainability, https://doi.org/10.1007/978-3-031-13516-3_8

production. For example, Texas—a Rule of Capture state—amended its constitution and later passed general legislation that allows for local and regional groundwater management to pursue sustainable production. In this case, state policy allows sustainability to be pursued at local and regional levels where decisions on sustainable management occur.

Without an existing supportive legal and policy framework, actors become critical (although even supportive legal doctrines and policies require the vigilance of actors to prevent doctrines and policies from changing in a way that harms sustainability). Having supportive actors can be difficult in aquifers not being pumped sustainably because sustainable management may work against the short-term economic benefits for existing groundwater producers (see Chap. 5). Furthermore, depending on the legal framework, policy changes may have to occur at the local, regional, state, and federal levels to support or even allow sustainable management. Changes at state and federal levels often do not occur unless severe consequences from groundwater production become evident.

Information and knowledge are also critical in following a pathway toward sustainable groundwater production. Aquifer monitoring and the resulting interpretations define the effects from groundwater production. Numerical groundwater flow models allow projection of future effects and estimates of sustainable levels of production. Educating groundwater users, the public, decisionmakers, and other stakeholders on the aquifer, the effects of production, and what the future may hold is important for supporting policy development. Along those lines, transparency, honesty, and relationships are critical, especially when detractors present alternative science that borders on the fantastical. In short, if there is resistance to managing sustainably, then there needs to be reliable information quantifying what is happening and what will happen if nothing is done.

Ideally, the evaluation of groundwater sustainability is adaptive, allowing a periodic review of unacceptable environmental, economic, or social consequences and the scientific understanding of the aquifer and how it responds to production.

In this chapter, I provide examples of different pathways to achieving sustainable groundwater production according to an identification of sustainability types identified by Mace (2021a). There are, of course, many examples informed by local political and hydrologic conditions. Because I wanted to include the flavor of history, local sentiments, and policy challenges, I have necessarily focused many examples on Texas where I am familiar—sometimes intimately—with the human and technical details of attaining sustainable production.

The literature is full of case studies in other countries. For example, Rinaudo and others (2020) discuss French, Australian, and Chilean sustainable groundwater management policies with lessons learned for other countries.

That's Unacceptable

A key part of achieving groundwater sustainability is identifying the unacceptable environmental, economic, or social consequences. As previously discussed, these are policy decisions, not scientific ones, although science provides knowledge to inform these decisions. Part of the governance of groundwater is: Who decides what is unacceptable? And what process do they follow to make these decisions?

In some cases, the decision was made—generally by a legislative body or a court—a long time ago to follow an early definition of safe yield. Sometimes the decision was to produce a safe yield defined by recharge (in violation of the Water Budget Myth). Other times safe yield was equated to the maximum amount that could be pumped regardless of effects. Yet other times, perhaps more appropriately, the decision was pushed down to local, regional, and/or state governmental administrative bodies. Here, recommendations may be made by technocrats to administrative decisionmakers with little public involvement.

Facilitated stakeholder dialogue should be used when planning for groundwater management (Smith and others 2016). Ideally, from an equity, change-management, and citizen-involvement viewpoint, decisionmakers should use facilitated stakeholder dialogues to help identify unacceptable environmental, economic, or social consequences.

Groundwater Sustainability Typologies

Mace (2021a) identified five types of groundwater sustainability: (1) hydrologically forced, (2) legislatively forced, (3) court forced, (4) desire driven, and (5) de facto. In some situations, hydrogeology is the primary driver for sustainable production—an aquifer can only produce what it can produce (Mace 2021a). At some point in the future, all aquifers will hydrologically force sustainable production (see next chapter). However, for some aquifers, this occurs on the order of years to decades while in others it requires decades to centuries.

For a legislatively forced system, a legislature, congress, or other deliberative (or apparently deliberative…) body requires the sustainable management of groundwater resources. The deliberative body may design a governance process through which unacceptable consequences are defined, may broadly define unacceptable consequences, or may distinctly define those consequences. Regardless of how and how detailed, the legal authority for sustainable groundwater management sources from the deliberative body. In court-forced systems, the legal doctrine, a court interpretation of the legislation, or some other court decision forces sustainable management and, perhaps, even defines unacceptable consequences of production. Court-forced sustainability may be a side effect of a court decision that doesn't directly address groundwater sustainability, such as the protection of ecosystems, endangered species, or private property rights.

Desire-driven sustainability is achieved through a local desire to manage groundwater sustainably without the direct legal or legislative forcing of regional, state, or national governments. This sustainability is generally achieved through local concerns over the impacts of unsustainable groundwater production such as on springs or groundwater-fed surface-water sources. de facto sustainability means that an aquifer is being produced sustainably with no human intervention. In other words, current groundwater production is below maximum sustainable pumping. This may be the case because the land surface is not amenable to irrigated agriculture (or any agriculture) and that the groundwater is too remote, at least at present, to affordably export it to a distant user. This may also be the case because the aquifer is not productive enough to support more than local household and livestock uses.

Subsequent sections provide examples of these different typologies and how groundwater sustainability can be achieved under them.

Hydrologically Forced: The Leon-Belding Aquifer in West Texas

Springs are valuable sources of water in arid regions, and West Texas is no exception.[1] Comanche Springs (Fig. 8.1) served as a source of water for Native Americans at least as far back as 20,000 years. Native American trails moved from one spring to the next, providing pathways for the "discovery" of foreign lands by European explorers followed by Comanches expanding through the plains and Americans seeking gold in California.

With the establishment of Fort Stockton at the springs, settlers began making homes in the immediate area, and some began to divert flows from the Comanche Springs—flowing at about 1850 liter per second at the time—for irrigation.

Starting in the late 1940s, several landowners in the Leon-Belding area began sinking wells and, upon finding ample water, started to irrigate with groundwater. As groundwater irrigation increased, springflow decreased at Comanche Springs, disastrously so in 1951 when springflows decreased from 1000 to 250 liters per second. In response, the spring irrigators sued the groundwater irrigators. However, because the legal structure for groundwater in Texas is based on the Rule of Capture (and there was not a groundwater conservation district at that time), the spring irrigators lost their court battle at the doorsteps of the Texas Supreme Court. Having lost the case, the springs stopped flowing in 1961, putting more than 100 spring-fed family farms out of business.

The Leon-Belding area overlies part of an aquifer the state recognizes as the Edwards-Trinity (Plateau) Aquifer; however, the area acts like an aquifer-within-an-aquifer with down-faulted blocks where permeability is enhanced through karst and where parts of the aquifer are locally

[1] This discussion is based on Mace and others (2020).

Fig. 8.1 Location of the Edwards, Edwards-Trinity, and Trinity aquifers in Texas showing Comanche Springs, the San Antonio and Barton Springs segments of the Edwards Aquifer, and Val Verde County

deeper than other nearby parts (Fig. 8.2). As groundwater pumping continued, wells in the shallower blocks saw their yields decrease over the irrigation season, dropping from 60 liters per second to 6 liters per second (Fig. 8.3). In addition, water levels in the aquifer started to not fully recover after each irrigation season, suggesting long-term nonsustainable production.

The end result was that the irrigators in the shallower blocks no longer had a reliable supply of groundwater and had to stop irrigating, putting them out of business. This resulted in reducing production back to

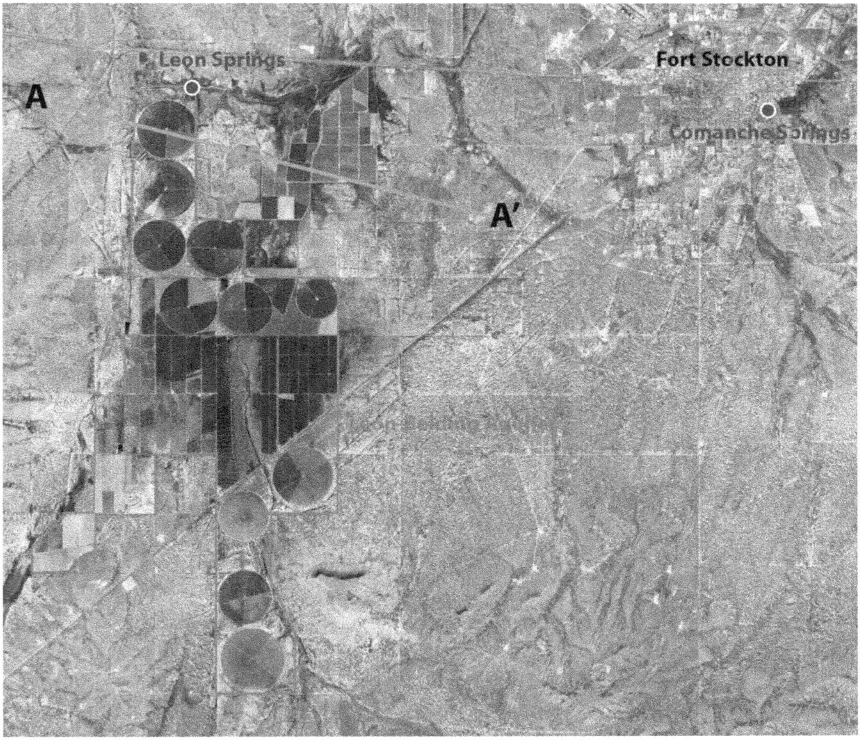

Fig. 8.2 Leon-Belding Aquifer in West Texas (base map from Google Earth)

sustainable levels. In other words, even in the absence of groundwater governance and management, production from the aquifer became sustainable due to hydrogeologic limits of the aquifer, and it happened over several years due to the relatively small size and high permeability of the aquifer and the relatively large amount of water produced from it.

Although the aquifer has been produced sustainably, groundwater production has created undesired consequences for some stakeholders. Comanche and Leon springs (and a number of other springs) went dry due to groundwater production, removing 2500 hectares of spring-irrigated agriculture. Ecosystems associated with those springs were destroyed, and the culture of Fort Stockton—long defined by the

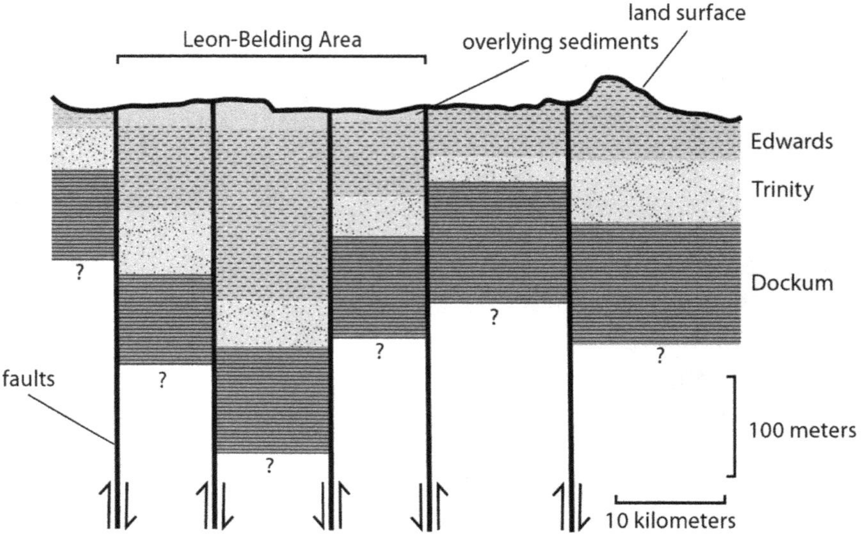

Fig. 8.3 Cross-section through the Leon-Belding Aquifer in West Texas (modified from Bumgarner and others 2012). Wells in the shallower part of the aquifer in the Edwards rocks (where the bottom of the Edwards is at a higher elevation) lost productivity or went dry when the aquifer was pumped beyond sustainable levels. Figure 8.2 shows the approximate location of the cross-section

springs—was affected. A nonsupporting legal doctrine and a lack of local governance and management doomed the springs.

Mace and others (2020) studied the history, hydrology, policy, and economics to assess the feasibility of bringing Comanche Springs back. Locals formed a groundwater conservation district in the county in 1999, and the district strives to manage the Leon-Belding area sustainably with tiered prior-appropriation permitting (export permits are proportionally reduced first followed by production permits [permits granted after historic and existing use permits were granted], followed by historic and existing use permits given at the time the district formed).

Although the district manages the aquifer sustainably, the sustainable level does not allow the springs to flow, and reducing permitted amounts almost guarantees a takings lawsuit (which alleges that the landowner is due compensation for the government's taking of private property). One way to bring springs back would be to purchase and retire existing permits; however, the district does not limit the number of permits, thus any purchased and retired permit would likely be greeted with a new permit (and new pumping).

A way to address the private property issue and yet allow for purchase and retirement of existing groundwater rights would be for the district to change to a correlative rights approach tied to subsurface geology similar to an approach used by Blumberg and Collins (2016) in the Carrizo-Wilcox Aquifer in Central Texas. Blumberg and Collins (2016) tied their correlative right not only to surface area over the aquifer but also net sand thickness in the aquifer. Although Blumberg and Collins (2016) do not use correlative rights to achieve sustainability, their approach could be used to do so assuming that stakeholders and the local regulator agree.

Besides the permitting challenge, there is also the challenge of existing rightsholders seeking to export groundwater to thirsty cities to the north (Midland, San Angelo, and Abilene) suffering from decreased surface-water reliability due to an extended drought in the American Southwest that may be a new normal due to climate change.

Legal Framework: Originally, no regulatory framework under the Rule of Capture. Later, it was the legislative creation and subsequent local confirmation of the local groundwater conservation district.

Actors and Policies: Originally, the actors for sustainable management were the irrigators that relied on springflow; however, the legal framework was not on their side (and they were not able to form a groundwater conservation district at that time). Later, it was the groundwater users, including those in a larger area than the footprint of the Leon-Belding Aquifer, and a local desire to manage sustainably to protect existing and historical use.

Information and Knowledge: Originally, little was known about the aquifer except springflows, where wells were, and how productive those wells were. As the human and legal drama unfolded, state and federal scientists collected more data on the aquifer to conclude that, indeed, groundwater pumping had dried up the springs; however, the information was too late to be considered in court proceedings (but, given that the Rule of Capture governed groundwater production at that time, the science probably would not have made a difference). Later, more science was developed with the financial support of the city, the groundwater conservation district, and a permittee seeking an export permit.

Why Is This Hydrologically Forced Sustainability? This aquifer was hydrologically forced into sustainable groundwater production because the aquifer could not support increased pumping, thus forcing less-advantageously located farms out of business, bringing production back into equilibrium with aquifer inflows. Later, in an example of desire-driven sustainability, locals created a groundwater conservation district to formalize sustainable development and protect historic use.

Hydrologically Forced (and Legislatively Nudged): The Great Artesian Basin of Australia

The Great Artesian Basin[2] hosts one of the largest aquifers in the world, spanning more than a fifth of the Australian continent (Fig. 8.4). Its characteristic mound springs—small hills of carbonate-cemented sand, silt, and clay at spring discharge points—have supported desert ecologies for eons and human life for millennia, upwards of 40,000 years (DEWHA 2009; DAWE 2020). The springs are important to aboriginal people as a source of water and as a critical part of their belief system (Mudd 2000; Harris 2002).

Aboriginal people dug the first wells in Australia (Powell and others 2015; Lawrence 2018) followed by European settlers, probably as early as 1606 at present-day Mapoon, Queensland, by Dutch sailors (Lawrence 2018). By the late 1800s, European settlers were drilling boreholes (bores in the local parlance) to tap shallow water tables and shallow confined aquifers that produced nonflowing artesian wells. Groundwater management (or lack thereof) started in Australia under the Rule of Capture (Barnett and others 2021). States and territories are charged with managing water by Australia's constitution (Turral and Fullagar 2007), but the federal government influences state and territory policy due to shared interests and federal financial incentives (McKay 2002).

[2] Dafny (2016) makes the (so far ignored) argument that the Great Artesian Basin is not, in fact, a basin.

Fig. 8.4 Location of the Great Artesian Basin in Australia (map generated by the author in ArcGIS and modified in Adobe Illustrator; sources: Esri, Airbus DS, USGS, NGA, NASA, CGIAR, N Robinson, NCEAS, NLS, OS, NMA, Geodatastyrelsen, Rijkswaterstaat, GSA, Geoland, FEMA, Intermap, HERE, Garmin, FAO, OpenStreetMap contributors, and the GIS user community)

David Brown[3] is credited with drilling the first flowing artesian well in the basin in 1879 near Wee Wattah Mud Spring about halfway between Bourke and Wilcannia and about 30 kilometers north of the Darling River (Cox 1895; Allan 1906; Habermehl 1980[4]). This well was only

[3] Brown supervised the drilling of the well.

[4] Many modern references, including this one, report a date of 1878 for the completion of the Kallara Station Well; however, references from the time report 1879 as the date.

12 meters deep, but the water rose 8 meters above the wellhead (Cox 1895) with a flow of 0.6 liters per second (Wilkinson 1882). Brown was not seeking artesian flow, so the well was a pleasant surprise to him, to scientists, and to government officials. In a bit of hydrogeologic foreshadowing, Wee Wattah Mud Spring, about 120 meters away from the well, stopped flowing after the well started flowing (Pittman and David 1903; Powell and others 2015).

Brown's well prompted the drilling of more exploratory wells including a deep bore in the Cunnamulla District in 1887—393 meters deep resulting in 3.5 liters per second of flow—which sparked a frenzy of well drilling (AWIC 1954). The number of wells drilled in Queensland alone rose from 34 in 1889 to 524 in 1899 (Nicol 2005). By 1915, there were over 1500 flowing artesian wells in the basin with a total flow of over 23,000 liters per second (Nicol 2005; Habermehl 2018). Blake and Cook (2006) refer to the period between 1880 and 1915 as the "Artesian Age" in the basin.

A typical well owner was a rancher who directed artesian flows into unlined canals (referred to locally as "bore drains"), some as long as 100 kilometers, for livestock and to irrigate pastures (Nicol 2005; Habermehl 2018; Fig. 8.5). Andrew "Banjo" Paterson colorfully captured the value of these free-flowing wells with the rhymes of "Song of the Artesian Water" published in 1896 (Paterson 1902).[5]

As early as 1891—and certainly by the early 1900s—well owners noticed that flows from their artesian wells had started to decline (Nicol 2005) and became concerned that they were exceeding the limits of the aquifer. In 1893, government scientists expressed concern about the number of wells, uncontrolled flows from the wells, and waste of the water (Williamson 2012). The scientists recommended that Queensland take control of artesian supplies, control drilling, and reduce waste (Williamson 2012). Queensland attempted to pass legislation soon thereafter to control groundwater use, but the effort failed due to landowner concerns over government regulation and a lack of knowledge on how serious the problem was (Williamson 2012; Habermehl 2018). The

[5] A favorite passage: "And it's time they heard us knocking on the roof of Satan's dwellin'; but we'll get artesian water if we cave the roof of hell in—Oh! We'll get artesian water deeper down."

Fig. 8.5 Historic postcard (circa 1910) of a flowing artesian well in Australia (photo by J. Beckwood, from the author's collection). Note that this well is controlled (it has a valve that allows the owner to control the flow, including turning the flow off) but that the canal is unlined

legislation returned in 1910 and passed, requiring permits for wells, drilling reports, construction requirements, and use reporting (Habermehl 2018) but not limitations on accessing and producing from the aquifer.

To address gaps in the scientific understanding of the aquifer, stakeholders, including scientists and engineers, began holding an Interstate Conference on Artesian Water, the first in 1912 with subsequent conferences held in 1914, 1921, 1924, and 1928 (AWIC 1945; Habermehl 2018). Researchers studied decreases in pressure and flows, the extent of the basin, the source and movement of groundwater, casing corrosion, and more efficient use (Habermehl 2018). The conferences resulted in systematic data collection and interpretation and the recognition of the need for controlled development (Barnett and others 2021). Despite a recommendation for controlled development, uncontrolled well drilling and groundwater production continued (Barnett and others 2021).

After a 1939 Interstate Conference on Water Conservation and Irrigation, Queensland formed the Artesian Water Investigation Committee (AWIC 1945; Williamson 2012). The Committee focused on summarizing the hydrogeology of the basin and "legislative factors related to the use and conservation of the water" (AWIC 1945). The Committee published an interim report in 1945 (AWIC 1945) and a final report in 1954 (AWIC 1954), the latter of which was timely because the demand for groundwater increased after World War II due to better drilling technology (Barnett and others 2021). By 1952, the number of artesian wells completed in the Great Artesian Basin was 2309 (AWIC 1954).

The Committee's 1954 majority report presented the most comprehensive understanding of the basin at that time (AWIC 1954; Habermehl 2018). The Committee identified water-level declines of up to 122 meters over 50 years and noted that several hundred wells had stopped flowing (AWIC 1954). They projected that another 660 wells would be expected to stop flowing in Queensland over the subsequent 50 to 60 years (AWIC 1954). However, the Committee also concluded that the aquifer would come into equilibrium and that about 500 wells "will continue to flow forever" at a total rate of 4800 liters per second, about half of the 10,000 liters per day flowing at that time (AWIC 1954). The Committee thought that groundwater production had increased recharge to the aquifer from about 2800 liters per second to about 5500 liters per second (although I suspect they were referring to effective recharge [which probably came at the expense of springs near the recharge zone]) (AWIC 1954).

The Committee saw the decrease in artesian flows as a natural consequence of producing water from the aquifer as the aquifer came into a new postproduction equilibrium. For economic reasons, they recommended against water conservation actions, arguing that it would simply delay the new equilibrium condition, although they acknowledged that conservation would allow greater use of the produced water (AWIC 1954). Although many wells would cease to flow, water would still be in the aquifer and accessible through pumps, with the Committee concluding that "flowing water is not essential" (AWIC 1954).

The recommendation that conservation was not necessary was not adopted by all members of the Committee. A thoroughly referenced minority report by a frustrated C.E. Parkinson challenged the majority's conclusion that conserving groundwater was not necessary, noting that 26.4 percent of flowing wells in Queensland had no control valves, 16.9 percent had deteriorated casing, 10.2 percent had leaking casing, 4.2 percent discharged to swamps, and 28.2 percent discharged to streams (Parkinson 1954; Williamson 2012). Although Parkinson lost his argument in committee, perhaps he took solace in Queensland passing legislation in 1954 requiring new artesian wells to be properly valved with their water distributed by pipe instead of open channel (Williamson 2012). However, older wells were allowed to flow as they had before although efforts began to rehabilitate older wells (Williamson 1996).

By 1963, there were about 3600 artesian wells in the Great Artesian (Fisher and Hahn 1963). Due to unsustainable production of artesian aquifers in the 1960s and 1970s, the federal government began to manage groundwater in a piecemeal fashion; however, a lack of management resources and some political interference resulted in permitting groundwater use above sustainable levels (Barnett and others 2021). To assess Australia's water resources, including groundwater, federal and state governments formed the Australian Water Resources Council in 1962 (Barnett and others 2021). The Council formed a Technical Committee on Underground Water, later renamed the Groundwater Committee and later, in 1985, formed the Great Artesian Basin Sub-Committee (Habermehl 2018). By 1977, the number of wells tapping into the basin had leveled off at about 1200 and flows from the aquifer had stabilized—and even recovered—at about 3500 liters per second (Williamson 1996).

In 1987, the Sub-Committee focused on the inefficient use of water for agriculture, particularly the use of unlined canals—which lost 95 percent of produced groundwater—and uncontrolled flows from wells (wells without valves that allowed the user to control the amount of flow from a well, including turning the flow off when not in use) and proposed the Great Artesian Basin Bore Rehabilitation Program (Woolley and others 1987; Barnett and others 2021). The Bore Rehabilitation Program started in 1989 and, by 1997, 40 percent of 1192 flowing artesian wells had been rehabilitated resulting in water savings of 1500 liters per second (Habermehl 2018).

After the release of the Brundtland Report in 1987 calling for sustainable development (WCED 1987) and the subsequent release of the working plan—Agenda 21—in 1992 to achieve sustainable development (United Nations 1992), Australia responded with the development of the National Strategy for Ecologically Sustainable Development in 1992 to provide equity between generations and maintain essential ecological processes (Turral and Fullagar 2007). A national policy for sustainability began with the 1994 National Water Reform Framework Agreement (CoAG 1994). This was notable since, previously, the states had the responsibility of managing water resources (Barnett and others 2021). Subsequent reports provided advice to the state and territories on groundwater management and sustainable yield (ARMCANZ 1996; Barnett and others 2021) and how to reduce production in the country's aquifers (NRMMC 2002). However, the ARMCANZ (1996) report defined sustainable yield in a way that allows groundwater mining (Turral and Fullagar 2007).

Stakeholders created the Great Artesian Basin Consultative Council in 1997, which comprised of federal, state, and local governments; ranchers; industry; land- and well owners; community and conservation groups; and aboriginal peoples (Habermehl 2018) and published a resource study of the basin in 1998 (Cox and Barron 1998). In 2000, the Council published the Great Artesian Basin Strategic Management Plan to address management issues, sustainable development, and water use with an implementation cost of $286 million (Australian) over the ensuring 15 years (GABCC 2000). In 1999, in response to the development of the management plan, the federal government and the states committed to the Great Artesian Basin Sustainability Initiative to bring uncontrolled wells under control and to convert dirt canals to pipes (DoE 2014). These programs provided incentives: a federal match to state dollars with some contribution from landowners (Nicol 2005).

By 2000, water levels in the aquifer had decreased by 80 meters and total flow from the Great Artesian Basin had decreased to about 17,000 liters per second despite more and more wells tapping into the aquifer (Nicol 2005). By 2000 in Queensland, 87 of 107 spring-groups mapped by 1898 had gone dry (Farifax and Fensham 2002). Nicol (2005) concluded that the flow decline was approaching equilibrium, but the data suggests otherwise, even suggesting a greater rate of decline in the decade leading up to 2000 (Fig. 8.6). There also wasn't consensus between agricultural and urban interests, environmentalists, economists, academia, and private interests on the best path to "water security" (McKay 2002). By the 2000s, water levels in the aquifer had decreased by up to 100 meters with the largest declines at the Queensland-New South Wales border (Habermehl 2018).

The Great Artesian Basin Strategic Management Plan and the Great Artesian Basin Sustainability Initiative were, in large part, a success, demonstrating the importance of having measurable goals. Between 1999 and 2017, federal and state governments invested $248 million (Australian) in the basin resulting in capping, rehabilitating, or decommissioning 750 wells and converting 21,390 kilometers of open canals to pipes, resulting in water savings of 7900 liters per second (DAWE 2020). The Great Artesian Basin Coordinating Committee replaced the Great Artesian Basin Consultative Council in 2004 (Habermehl 2018).

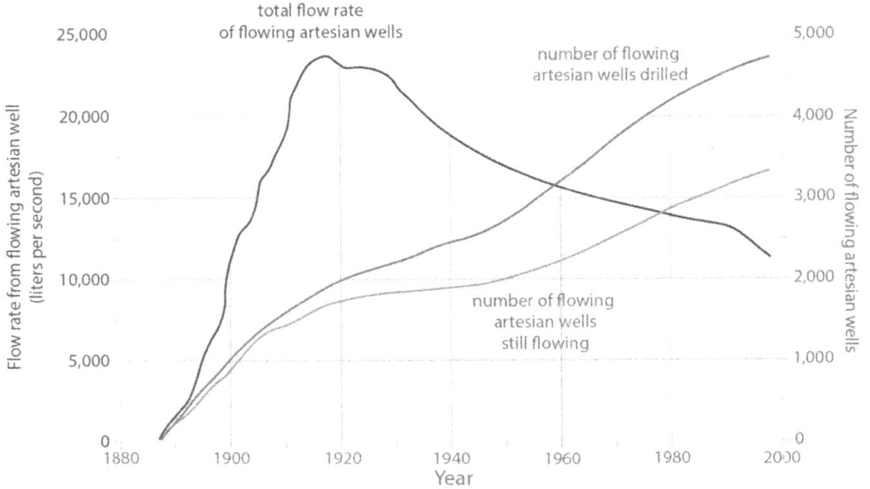

Fig. 8.6 Number of flowing artesian wells drilled and still flowing and total flow from flowing artesian wells from 1880 to 2000 in the Great Artesian Basin of Australia (after Habermehl 2018)

The Australian Department of Agriculture, Water, and the Environment, working with the Council and other stakeholders, published an updated plan in 2020 (DAWE 2020). Instead of recommendations, the plan identifies "opportunities," including

- increasing stakeholder awareness that the aquifer is a declining and finite resource;
- developing frameworks to manage third-party impacts;
- developing adaptive management approaches to accommodate increases and decreases in artesian pressures;
- reviewing maintenance and decommissioning practices for wells and canals;
- improving cross-border management of groundwater resources;
- examining impacts of emerging and new water users on water consumption, quality, and pressure;
- tracking of water use, water levels, and flows to inform decision-making, particularly on climate change and new, emerging, or expanding uses;
- recognizing and conserving biodiversity resources and cultural heritage values that rely on water;
- clearly defining rights and responsibilities of water users;
- promoting the timely and robust collection of information for use as the basis for decision-making;
- understanding and accounting for the effects of land use change on the recharge zone;
- accounting for production-induced impacts between artesian and nonartesian aquifers;
- assessing and responding to the effects of climate change on long-term management of the aquifer;
- supporting the development of skills and knowledge in the community to provide well construction and maintenance, water delivery infrastructure design, and asset condition monitoring activities;
- assessing changing populations and related economies over the aquifer;
- recognizing the need to maintain ongoing urban water infrastructure over the aquifer;

- establishing a proactive aquifer-wide governance structure and process to maintain the profile of the aquifer and ensure effectiveness of the evidence-based risk management approach;
- implementing compatible policy approaches to strengthen protection of springs and groundwater-dependent ecosystems;
- developing and implementing innovative water-use efficiencies in all sectors through improved technologies and better water management practices;
- enabling and encouraging the beneficial use of coproduced water; and
- investigating a long-term funding arrangement for aquifer-wide replacement of wells.

This plan also identified emerging challenges including gas injection, enhanced recharge, large well fields, unconventional gas extraction, and geothermal energy (DAWE 2020).

Today, about 3100 controlled and uncontrolled wells flow about 17,400 liter per second (Habermehl 2018), which represents an uptick in use mostly due to water use associated with coal gas production, which increased from less than 160 liters per second in 2005 to 570 liters per second in 2013 to 2200 liters per second in 2016 (Frontier Economics 2016). Although most flowing artesian wells are now under control, there still remain at least 430 uncontrolled wells (Brake 2020). Most springflows have declined and, in some areas, springs have gone dry (Habermehl 2018). By 2005, 56 percent of 171 spring complexes in the Great Artesian Basin had gone dry (Fensham and Fairfax 2005).

Legal Framework: Originally, there was no regulatory framework under the Rule of Capture. Later, well completion and water conveyance requirements were legislated for new wells. Federal and state governments have incentivized the retrofitting of older wells (with valves to bring uncontrolled artesian flows under control) and unlined canals (to move water through pipes).

Actors and Policies: Originally, government scientists called for state control of groundwater, but landowners disagreed. Later, stakeholder groups of various public and private entities came together to identify goals to achieve sustainable management which were implemented through federal and state incentives to well owners.

Information and Knowledge: Because of early concerns about the long-term viability of the aquifer, some of which were caused by scientific disagreements on the source of water to the aquifer, a great deal of science was developed early and then nearly continuously for the past 120 years. The science informed—and continues to inform—policy decisions in the aquifer.

Why Is This Hydrologically Forced and Legislatively Nudged Sustainability? This aquifer was hydrologically forced into sustainable groundwater production because, as production increased through more wells, the flows declined as water levels declined. As long as groundwater is predominantly produced through flowing artesian wells, the system will come into a new, sustainable equilibrium, albeit with negative consequences for springflows and pumpless groundwater production. Legislative action has increased the chances of achieving sustainability by increasing the efficiency of groundwater production and use through well completion and canal design requirements that applied to new wells drilled after 1954 and through incentives for landowners to bring flowing artesian wells under control and to convert open canals to pipes for conveying water. Without direct controls over production amounts and an overall cap on production, the sustainability of the system is, in my opinion, tenuous. Although the decrease in flows from artesian wells is a natural tapering of the supply through unnatural penetrations, pumps can easily circumvent flow reductions.

Legislatively Forced: California's Sustainable Groundwater Management Act

Groundwater in California is primarily sourced from its 515 alluvial groundwater basins (CDWR 2021a) nestled between mountains in the Basin and Range province of the western United States. Groundwater provides about 38 percent of the state's total water supply (CDWR 2021a). In the beginning, up until 1949, California had no regulatory or legal framework for groundwater, resulting in, essentially, de facto Rule of Capture. However, in a legal case between the City of Pasadena and the City of Alhambra,[6] a California Supreme Court ruling resulted in the court's adjudication of groundwater in the basin achieved through assessing existing production and proportionally reducing that production to the safe yield. Although local adjudications were now possible, statewide management eluded the state despite various attempts due to concerns over economic impacts (Sax 2002; Leahy 2016; Milman and Kiparsky 2020). The end result was de facto Rule of Capture until someone sued, upon which the court adjudicated groundwater rights to the safe yield.

After two droughts in the 2010s, Governor Jerry Brown, citing climate change, drought, and population growth, tasked the state's water agencies (the California Natural Resources Agency, the California Environmental Protection Agency, and the California Department of Food and Agriculture) with identifying key actions over the subsequent five years to "provide for the foundation for sustainable management of California's water resources" (Leahy 2016). Meanwhile, the State Water Resources Control Board was developing a concept paper on groundwater management (SWRCB 2013; Leahy 2016). The concept paper—published on October 4, 2013—called for (1) sustainable thresholds for water-level declines and water quality, (2) water quality and water level monitoring and assessment, (3) governance structures and management to prevent overproduction and achieve sustainability goals, (4) funding, and (5) oversight and enforcement (SWRCB 2013). On October 31, 2013, the state's water agencies published a draft of the California Water Action Plan (CNRA and others 2013). The plan called for (1) recharging aquifers through injection or infiltration basins, (2) conjunctive management with surface water, (3) management to reverse declining water levels and manage groundwater sustainably, and (4) data collection and scientific studies (CNRA and others 2014).

On January 9, 2014, the Governor released his proposed budget, which included $619 million to implement the Water Action Plan, followed a little more than a week later with a drought proclamation (Leahy 2016). While developing bill language, there were numerous hearings, public meetings, third-party reports, and lobbying concerning sustainable groundwater management. The Governor's Office held its own

[6] City of Pasadena v. City of Alhambra, 207 P.2d 17, 23 (Cal. 1949).

stakeholder meetings and even posted its own draft statutory language on its website (Leahy 2016).

The final bill restricted its jurisdiction to the 127 basins defined by the Department of Water Resources as high and medium priorities based on population, population growth, and number of public water-supply wells, irrigated acreage, and relative reliance (Leahy 2016). Although a fraction of the 515 total basins in the state, these 127 basins held 96 percent of California's groundwater production (Leahy 2016). Also exempted from the Act were the 26 court-adjudicated basins and 3 in litigation among the 127 high- and medium-priority basins.

During consideration of the bill, there was confusion as to whether basins should be managed to safe yield or sustainable yield with the bill's authors referring to a paper by Rudestam and Langridge (2014) (Leahy 2016). The Act ultimately defined sustainable groundwater management as "…the management and use of groundwater in a manner that can be maintained during the planning and implementation horizon without causing undesirable results" (Dickinson and others 2014; Pavley and others 2014). The Act defined undesirable results as chronic water-level declines, significant and unreasonable reduction of groundwater storage, significant and unreasonable seawater intrusion, significant and unreasonable land subsidence that interferes with surface land uses, and depletions of interconnected surface water that have significant and unreasonable adverse impacts on beneficial uses of the surface water (Dickinson and others 2014; Pavley and others 2014). The Act defined sustainable yield as "…the maximum quantity of water, calculated over a base period representative of long-term conditions in the basin and including any temporary surplus, that can be withdrawn annually from a groundwater supply without causing an undesirable result" (Dickinson and others 2014; Pavley and others 2014).

The Act required local agencies in a groundwater basin to form groundwater sustainability agencies to develop groundwater sustainability plans that would achieve "groundwater reliability" by 2040 or 2042 (CDWR 2020a, 2021b). The Act required the California Department of Water Resources to evaluate the groundwater sustainability plans to assess that they are based on the best available science and are reasonably likely to achieve the identified sustainability goals (CDWR 2021b). Alternatives to the groundwater sustainability plans are allowed for existing groundwater management plans, adjudicated basins, and basins that have been produced below their sustainable yield for at least ten years. Users in basins that refuse to participate in planning will then have plans developed by the state (a strong incentive for locals to develop their own plans).

For critically overdrafted basins,[7] groundwater sustainability agencies were required to submit their plans in January 2020 while those of high and medium priority were required to submit their plans in January 2022 (CDWR 2020a). Submitted plans are currently under review and available for viewing at the California Department of Water Resources website (https://sgma.water.ca.gov/portal/#alt).

There are still potential challenges to groundwater sustainability in California, besides the time frame. One is a collision with private property rights (Griggs 2021), although adjudication to safe yield has survived this collision. Other areas of concern include the limitation to alluvial basins (leaving underlying and bordering hard-rock aquifers unregulated), leaving underlying brackish groundwater unregulated, and leaving too many basins at low-priority (and therefore unregulated) (Thompson and others 2021). In my opinion, a time frame over which groundwater production is sustainable is problematic for achieving indefinite sustainability.

Legal Framework: State legislative action prompted by the governor and taken in response to several droughts and concerns over declining groundwater levels; state agency oversight, regional planning and regulatory action, and local water district implementation.

Actors and Policies: The governor, the legislature establishing a desire to manage aquifers sustainably and broadly establishing unacceptable consequences, state agencies coordinating and policing implementation of the act, groundwater sustainability agencies establishing the details of unacceptable consequences and actions to achieve sustainability, and the member agencies of the groundwater sustainability agencies.

---

[7] The Sustainable Groundwater Management Act defines a critically overdrafted basin "when continuation of present water management practices would probably result in significant adverse overdraft-related environmental, social, or economic impacts."

Information and Knowledge: Science on the many basin aquifers in the state varies considerably. Many basins do not have numerical groundwater flow models.

Why Is This Legislatively Forced Sustainability? Management of these high- and medium-priority basin aquifers are an example of legislatively forced sustainability because the state legislature required active sustainable management of the basins. Before the Sustainable Groundwater Management Act passed, some basins were court-adjudicated to safe yields and thus represent court-forced sustainability.

Legislatively Forced: Namibia, Africa

Outside of the Saharan Desert, the Republic of Namibia (Fig. 8.7) is the driest country in Africa and one of the driest in the world, where 97 percent of rainfall evapotranspirates, 2 percent runs into rivers, and 1 percent percolates into aquifers (Heyns and others 1998). Perennial rivers flow at the northern and southern boundaries, but the interior of the country—bracketed by the Namib Desert in the East and the Kalahari in the West—relies a great deal on groundwater (Schneider and others 2001; Heyns and Struckmeier 2001).

In such a dry country, it's not surprising that water is an important part of culture. Springs provide water to flora and fauna, and the San, Damara, and Nama people have lived in the area since prehistoric times (Wallace 2011) for at least 10,000 years (Jokisch and others 2018). Local people also constructed hand-dug wells to enhance supplies (Kluge and Polak 2018; Christelis and Struckmeier 2001). Several towns take their name from historical springs, such as Gobabis, a spring named "place of the elephants" (Angula 2001). Legend has it that the Nama established Hoachanas, one of the oldest settlements in southern Namibia, when the Khai||khaun's hunting dogs discovered a spring there in 1695 which was named Arigu-|aus (Fountain of the Dogs) (Umenne and Chirwa 2012; Hinz 2016; Fig. 8.8). A number of towns owe their existence to the springs they were founded at, including the capital, Windhoek (Heyns and Struckmeier 2001); Karasburg, founded at Nomsoros (water between the limestones); and Warmbad founded at Gei-ous (big hot water source) (Angula 2001).

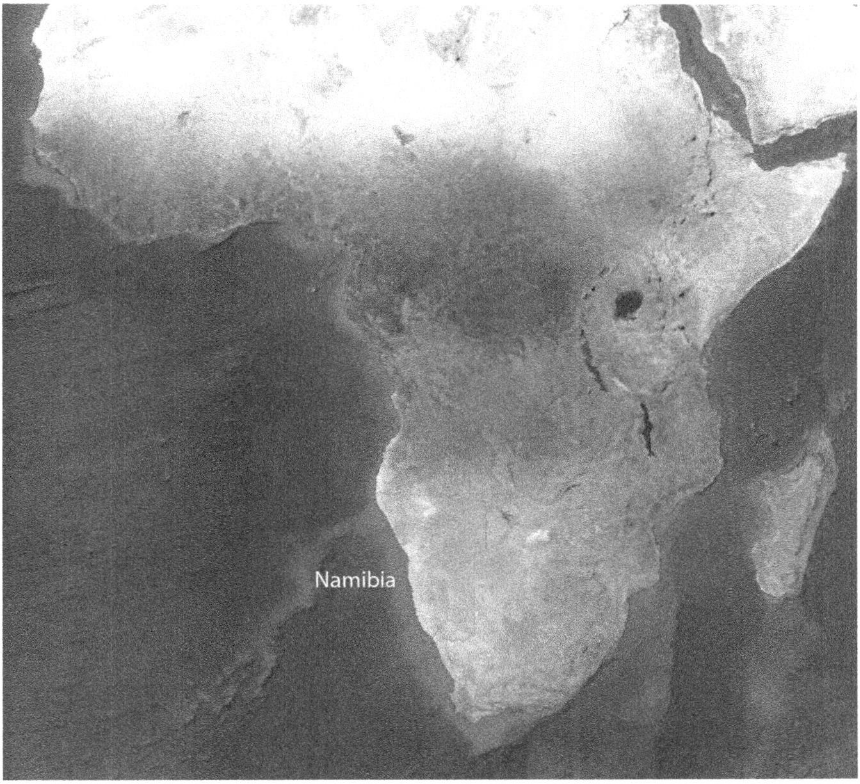

Fig. 8.7 Location of Namibia (map generated by the author in ArcGIS and modified in Adobe Illustrator; sources: Esri, Maxar, Earthstar Geographics, CNES/Airbus, DS, USDA, USGS, AeroGRID, IGN, and the GIS user community)

The Portuguese were the first Europeans to explore the coast in 1485 (Ravenstein 1900), but the area did not come under European colonial rule until 1884 when Germany claimed it as a protectorate (Wallace 2011). The first drill rig appeared in the area in 1903 to research groundwater resources in the country (Heyns and Struckmeier 2001).

During World War I, South Africa invaded the area, deposed German rule, and occupied the territory (Wallace 2011). Known as South West Africa at the time, South Africa agreed to administer the area until the residents were ready for self-governance. After World War I, the Irrigation

Fig. 8.8 Arigu-|aus (Fountain of the Dogs) in Hoachanas, Namibia (photo by Rainer Hog of Bitterwasser (PTY) Ltd., used with permission)

Department for the territory created the Boring Division[8] to find groundwater for livestock, farming, and irrigation (Heyns and Struckmeier 2001).

South Africa drug its feet on allowing self-determination and, in response to a United Nations decree in the 1940s that would have led to

[8] Take that, Elon Musk!

the independence of South West Africa, South Africa began actions to annex the territory. In 1968, after a lengthy international court process, the United Nations declared South Africa's actions as illegal and, following the wishes of the locals, renamed the disputed territory Namibia. By 1990, after decades of conflict, Namibia finally won its independence from South Africa and began its own path as a parliamentary democracy. Namibia's relatively recent independence and entry onto the path of self-determination allowed it to develop a constitution, laws, and policies with the benefit of a long history of past policy experience from across the world.

Similar to other countries, Namibia saw an explosion in the number of wells once deeper wells became more affordable. Ranchers relied on springs and surface water until the mid-1900s when they started to drill wells, increasing tenfold between 1960 and 1990 (Bollig 2013). By 2000, more than 100,000 wells had been drilled with about half still in production (Heyns and Struckmeier 2001). As of 2020, about 45 percent of water for towns and farms sources from groundwater (Heyns and Struckmeier 2001). In some parts of Namibia (the Stampriet Basin, the Malta höhe-, and the Oshivelo area), landowners can drill and have flowing artesian wells (Christelis and others 2001). Hand-dug wells also remain an important source of water (Kluge and Polak 2018).

Namibia, like many countries in Africa (Gaye and Tindimugaya 2019), has not studied its aquifers in detail. For example, monitoring data on Namibia's aquifers is typically less than 50 years, and the lack of information on their aquifers makes it challenging to assess the sustainable yields (Heyns and Struckmeier 2001). In 2001, Namibia worked with Federal Republic of Germany to map out potential groundwater resources in the country (Christelis and Struckmeier 2001). This survey found that 42 percent of the country is underlain by aquifers of which 62 percent are porous (unfractured) aquifers and 38 percent are fractured (Christelis and Struckmeier 2001). Scientists estimate the combined long-term sustainable yield of Namibia's aquifers at about 9500 liters per second (Christelis and Struckmeier 2001); however, it appears this number is based on the total amount of recharge and would be lower if springs and other discharges are maintained. By 2000, many of the country's alluvial aquifers were being pumped unsustainably (MAWRD 2000).

The Republic of Namibia approved its constitution in 1990 and states that it owns all water in the country, including groundwater (Constitute 2021, Article 100). Heyns and Struckmeier (2001) state that "[e]quitable and sustainable development, as spelled out in the Constitution of Namibia, includes the proper and wise use of all natural resources including ground-water, to supply the water needed without harming the environment." However, Namibia's constitution declares that it will adopt policies aimed at the "…maintenance of ecosystems, essential ecological processes and biological diversity of Namibia and utilization of living natural resources on a sustainable basis for the benefit of all Namibians, both present and future…" (Constitute 2021, Article 95, Section l). Living natural resources generally include plants, animals, and microorganisms and not minerals and water. While water resources, including groundwater, are not directly protected under this language, their protection is generally, but not absolutely, required to protect living natural resources. Regardless, Namibia was the first African country to include protection of the environment in its constitution (Nhuleipo 2022).

Namibia's water administration was influenced by the concept of Integrated Water Resource Management (Kluge and Polak 2018), which integrates demand-side water management (water conservation and efficiency) along with water-supply development. The Global Water Partnership defines Integrated Water Resource Management as "the coordinated development and management of water, land and related resources by maximising economic and social welfare without compromising the sustainability of vital ecosystems" (Solanes and Gonzalez-Villarreal 1999). In 2000, the government published a white paper on developing a national water policy that uses "…an integrated, multisectoral framework that considers issues of decentralization, social equity, ecological protection, and economic growth" (MAWRD 2000). Namibia's implementation of Integrated Water Resource Management can be regarded as a success due to the much greater access to safe drinking water than during the apartheid system of South Africa (Polak and Kluge 2018), but others have criticized the implementation (e.g., Remmert 2016).

The Water Resource Management Act No. 24 of 2004 required permits to use groundwater and included "management of water resources so as to promote sustainable development" as a fundamental principle. The Act also established basin management committees that, among other duties, were to "monitor and report on the effectiveness of policies and action in achieving sustainable management of water resources in its water management area." The act also gave the minister responsible for water the power to determine a safe yield to inform aquifer management and require an aquifer to be used sustainably, including requiring the reduction of use below the safe yield. The Act defined safe yield as "…the amount of water which may be abstracted from an aquifer at a rate that will not reduce the supply to such an extent as would render such abstraction harmful to the aquifer, quality of the water or environment."

The Water Resources Management Act No. 11 of 2013 superseded the Water Resource Management Act No. 24 of 2004. It kept permitting requirements but changed the fundamental principle to promote "…the sustainable development of water resources based on an integrated water resources management plan which incorporates social, technical, economic, and environmental issues…." Basin management committees remained. The Act uses the term "groundwater potential" instead of safe yield, defining it as "…the volume of water which can be abstracted over a period from an aquifer without adversely affecting the quantity and quality of the water or the environment above the aquifer…." The minister over water retained their ability to force users to produce less than the groundwater potential.

While under South African rule, local water-supply administration and costs were borne by the state; once Namibia became independent, the state pursued a decentralized model, pushing administration (and costs) to locally controlled water committees (Schnegg and others 2016). While this approach is giving stakeholders direct control over management of their water resources, it has also pushed costs down to people that may not be able to afford the costs (Falk and others 2009; Schnegg and others 2016). Furthermore, community-based management has challenged locals, who "…are more interested in having a reliable and affordable water supply than having a say in water management" (Polak 2014). Namibia has recognized the financial issues since 2000, but—as of 2016—had not addressed them (Schnegg and others 2016).

Each major aquifer has an aquifer management plan that includes aquifer monitoring and the employment of management strategies to ensure sustainable use (Heyns and Struckmeier 2001). Namibia has been storing surface water underground and increasing recharge through slowing surface water (to increase infiltration) and the use of infiltration basins (Christelis and others 2001).

Windhoek is a good example of groundwater exploitation and subsequent integrated water management. Windhoek was established due to the hot springs along the Klein Windhoek River (van Rensburg 2018) that issued from the Kuiseb Schist, known locally as the Windhoek Schist (Mapani 2005). As it grew and needed more water, wells were drilled at the springs, which caused them to go dry in the 1920s (Christelis and others 2001). By the time the reservoir behind the Goreangab Dam was completed, groundwater was being produced from the local aquifer at more than twice the sustainable level (van Rensburg 2018).

By 2000, the wellfield had 50 wells and provided about 10 percent of the city's water supply as compared to 80 percent sourced from surface-water reservoirs and 10 percent from a direct potable reuse project (Christelis and others 2001). Windhoek is also injecting surface water into the local aquifer, depleted of about 50 billion liters from groundwater production, to address evaporative losses from surface-water reservoirs (Christelis and others 2001) and provide a buffer from droughts (van Rensburg 2018). Windhoek has also improved water conservation and efficiency such that it used the same amount of water in 1997 as compared to 1990 despite a 45 percent increase in population (Christelis and others 2001).

A drought from 2014 to 2016 exhausted Windhoek's surface-water supplies, forcing it to depend on water reuse, demand management, and its managed aquifer recharge program to avoid running out of water (van Rensburg 2018).

Namibia is expected to warm by about 2 °C by the end of the century (Jokisch and others 2018), and climate change is already affecting spring discharge (Soetaert and others 2022). Groundwater resources in drier parts of Africa, which are already marginal, are expected to be more effected by climate change than wetter areas (Macdonald and others 2009). Windhoek is expected to warm by 1 to 4 °C with a concomitant

20 percent decrease in rainfall and a 10 to 15 percent increase in evaporation (van Rensburg 2018). Climate change and population growth will continue to challenge Windhoek's water resources.

Many African countries do not have policy and governance in place to manage groundwater resources but still have an opportunity to manage their resources sustainably (Gaye and Tindimugaya 2019). Although imperfect, Namibia may be a good example for the rest of Africa as well as the world.

Legal Framework: The sustainability of living natural resources, which rely on water, is included in the constitution. Legislation encourages sustainable development of water resources and gives the federal government the ability to force aquifer management to sustainable levels if locals are not able to.

Actors and Policies: Water governance includes federal, regional, and local governing bodies.

Information and Knowledge: Information and knowledge is growing in the country and aquifer management plans include plans for data collection.

Why Is This Legislatively Forced Sustainability? Legislation establishes sustainability as the goal, gives the federal government the ability to force aquifer regulation to sustainability, and gives locals the power to locally manage their groundwater resources.

Court Forced: The San Antonio Segment of the Edwards Aquifer

The Edwards Aquifer of Central and South-Central Texas is an amazing karst groundwater system, capturing water funneled from the Texas Hill Country and recharging it into the aquifer through losing streams. The San Antonio Segment, which is about 300 kilometers long and 65 kilometers wide, is the primary source of water to San Antonio, and supports irrigation for farms for 120 kilometers west of the city (Fig. 8.1). The state's largest and second-largest spring systems issue from the aquifer, and downstream users rely on the water.

Unlike many karst systems, it is easy to drill a well into the confined part of the Edwards Aquifer and be rewarded with amazing amounts of water. It is not uncommon to produce 60 liters per second from an Edwards well and see no measurable drawdown (Hovorka and others 1998). As a result of this amazing productivity, San Antonio—and other communities—transferred from springs to wells to meet their needs, and farmers ironed out the wrinkles in Mother Nature's fickleness with wells.

Although the overall sustainable production from the San Antonio Segment of the Edwards Aquifer wasn't threatened until the 1990s, local production in the 1900s had already turned the previously reliable San Antonio Springs and nearby San Pedro Springs in San Antonio into ephemeral features (Mace 2021a).

Desert springs are known for the development of unique ecosystems, and the springs of the Edwards Aquifer are no different. Although it is unknown what unique species resided in San Antonio and San Pedro springs before they went dry, Comal and San Marcos springs—the remaining major discharge points for the Edwards Aquifer (with Comal Springs being the largest spring in the American Southwest)—hold several unique species of plants, fish, and insects. While San Marcos Springs has never gone dry in recorded history, Comal Springs stopped flowing (although didn't go completely dry) in 1956 during the drought of record with drought-induced increased pumping.

Concerns about springflow and plans by the Nueces River Conservation and Reclamation District to build concrete-lined canals to avoid losing water to the aquifer led to the creation of the Edwards Underground Water District in 1959 (EUWD 1989). The district initially focused on understanding the aquifer and increasing recharge and broadened to include water conservation and assisting users to find alternative sources of water. Originally comprised of five counties, the western two rural counties dominated by irrigation use left the district in 1989 as discussions turned to decreasing use during drought.

The United States Congress passed the Endangered Species Preservation Act in 1966 to list native animal species as endangered and giving them limited protections (USFWS 2021). The Endangered Species Act in 1973 extended protection to plants and invertebrates and gave all species expanded protections (USFWS 2021). By 1990, several species in Comal and San Marcos springs had been listed as endangered by the federal government.

In 1991, the Sierra Club—an environmental nongovernmental organization—and a professor with the University of Texas at Austin filed a lawsuit in United States district court against the Secretary of the Interior and United States Fish and Wildlife Service alleging that the secretary had allowed damage to the endangered species by not ensuring water levels in the aquifer adequate to sustain springflows at Comal and San Marcos springs (Votteler 1998). In February 1993, a federal judge ruled for the plaintiffs and required the Service to determine springflow requirements to protect the species and for the state to prepare a plan to protect minimum aquifer levels and springflows lest "…the 'blunt axes' of Federal intervention have to be dropped" (Votteler 1998).

Under the threat of federal control, the Texas Legislature responded in May 1993 with the passage of legislation that abolished the Edwards Underground Water District and replaced it with the more powerful Edwards Aquifer Authority. After several more lawsuits concerning voting rights, agricultural activities, constitutionality, and damage to the species during another drought and subsequent legislative adjustments to the Edwards Aquifer Authority Act, the Authority finally assumed its role as the aquifer's regulator in 1996 (Votteler 1998).

In this case, unrelated Federal regulation[9] led to the sustainable management of the aquifer through court action and threatened federal control. A subsequent stakeholder-process, mandated by later state law, developed a habitat conservation plan to identify through best available science what the minimum flows needed to be and what management activities, such as water conservation and aquifer storage and recovery (excess Edwards water stored in the Carrizo Aquifer to the south), needed to occur for springflows to be maintained during a repeat of the drought of record (Gulley 2015). Interestingly, one alternative source of water to the Edwards Aquifer for San Antonio is the presumably unsustainable production of groundwater from the Wilcox Aquifer 225 kilometers to the northeast of San Antonio.

---

[9] By "unrelated," I mean that the Federal regulations were about endangered species, not about sustainable groundwater management.

To date, the Edwards Aquifer Authority continues to operate despite additional lawsuits working at the fringe of private property rights and central regulation. At the moment, stakeholders and the Authority are working on a new habitat conservation plan that will include an updated numerical groundwater flow model and projected impacts of climate change.

Legal Framework: The Endangered Species Act at the federal level, legislative action at the state level, regulatory action at the regional level.

Actors and Policies: The federal government setting the unacceptable consequence that defined groundwater sustainability (damage to endangered species that relied on springflows to maintain their habitat and thus their continued existence), the state legislature that created the regional authority to regulate groundwater production to avoid damage to the endangered species, federal and state courts that clarified what the regional authority could and could not do and how it would be governed, and the regional authority (with an elected board) that worked with scientists, permitholders, and other stakeholders to refine springflow requirements, pumping restrictions, and strategies to avoid the unacceptable consequence.

Information and Knowledge: The Edwards Aquifer is one of the most studied aquifers in the state and had several models available for use to assess how production and permit reductions during drought would affect springflow. Since the advent of stakeholder-driven regional water planning in 1997, Texans expect water decisions to be transparent and include robust stakeholder involvement (Mace 2021b). Due to the long and storied history of groundwater issues in the area, the importance of (still-flowing) springs to local economies, and the responsive nature of the aquifer (the San Antonio newspaper reports the aquifer level every day on the front page), most citizens are aware of the importance of the aquifer.

Why Is This Court-Forced Sustainability? This aquifer is an example of court-forced sustainability because the federal court forced sustainable groundwater management to the benefit of endangered species.

Legislatively Forced: The Albian Aquifer of the Paris Basin, France

After Georges Mulot drilled his famous well in Grenelle outside of then Paris between 1833 and 1841 (described in Chap. 3), people began drilling more and more wells to tap into the underlying Albian Greensands Aquifer in the Paris Basin (Rinaudo 2020; also referred to as the Albian Sands (Barraqué and de Marsily 2010) and the Albien green sands (Maréchal and Rouillard 2020), Fig. 8.9). As industry and others drilled

Fig. 8.9 Extent and formation temperatures of the Albian Aquifer in France (map generated by the author in ArcGIS and modified in Adobe Illustrator; sources: Esri, Airbus DS, USGS, NGA, CGIAR, N Robinson, NCEAS, NLS, OS, NMA, Geodatastyrelsen, Rijkwaterstaat, GSA, Geoland, FEMA, and the GIS user community; aquifer data from MetteSvaneDK)

into the confined Albian Greensands, water levels dropped in the aquifer (Dupaigne and Violette 2020; Rinaudo 2020).

With an elevation of 51 meters above sea level at the Grenelle Well site and predevelopment water levels in the Albian Greensands Aquifer at about 123 meters (water levels from Dupaigne and Violette 2020), it is no wonder that Grenelle Well made such an impression to the public (and the world) when it flowed with about 72 meters of artesian pressure at land surface. To celebrate the flows, the city built a 43-meter tall, cast iron fountain nearby to show the flows. However, four months after the well came in, the flow stopped and then returned but with a lower flow and carrying silty sands, probably due to a poor well completion that allowed sediment to sluff into the borehole (Meunier 1978).

In 1855, the City of Paris decided to drill a well in Passy, about 6.5 kilometers to the northwest of the Grenelle Well. At a depth of 548 meters, the Passy Well achieved a flow of about 200 liters per second (Meunier 1978). However, once the Passy Well began to flow, the flow at the Grenelle Well declined from 11 to 7 liters per second (Meunier 1978), a textbook case of well interference. By 1904, flows had presumably decreased enough such that the immense fountain built on Esplanade Jacques Chaban-Delmas (Fig. 8.10) no longer flowed and was replaced by a statue of Louis Pasteur.[10] By 1910, the Grenelle Well had stopped flowing and, by 1930, the Passy Well had also stopped flowing (Meunier 1978). The introduction of rotary well drilling in the early 1930s sparked more well construction, increasing well production from about 10 million cubic meters per year to about 35 million cubic meters per year (Contoux and others 2013). By 1935, water levels in the Albian Sands Aquifer in the Grenelle area had dropped to about 50 meters above sea level—a 73-meter drop from predevelopment conditions.

The foundational law in France traces back to Roman times through *res nullius* (Rinaudo 2020), literally "nobody's thing." The Rule of Capture, discussed in Chap. 7, traces back to this Roman concept as well (Blumm and Ritchie 2005). *Res nullius* formally became part of the civil code in France in 1804 and remains in the law today (Rinaudo 2020).

[10] It's unclear when the flows in the well were low enough that the fountain was turned off. A photograph taken in 1880 does not show the fountain flowing.

Fig. 8.10 Fountain built for the flowing artesian well in Grenelle, France (scanned image of a postcard in the author's collection)

The French government saw the rapid drop in water levels in the Paris Basin as a threat to France and the franc, which resulted in the passage of a law—the 1935 Water Act—requiring permission from the Mining Administration to drill a well deeper than 80 meters (Barraqué and de Marsily 2010).

The 1935 Water Act originally only applied to the area around Paris but was later extended to other deep basins (Barraqué and de Marsily 2010). The Act effectively stopped the drilling of deep wells in the Paris Basin (Barraqué and de Marsily 2010) while allowing existing users to continue with their nontransferable use of the aquifer (Rinaudo 2020). Groundwater production in the Paris Basin declined to less than 25 million cubic meters per year by the mid-1940s and about 20 million cubic meters per year by 1960 (Contoux and others 2013). However, water levels in the aquifer continued to decline over this period (Contoux and others 2013).

In the late 1950s, the French government established a water-level monitoring network in the Paris Basin and encouraged the development of groundwater models, the first of which had been developed by the late 1960s (Rinaudo 2020). By the mid-1960s, water levels at the Grenelle Well had dropped another 25 meters to about 25 meters above sea level (Dupaigne and Violette 2020). In 1964, the French government considered changing the legal status of groundwater but instead, through the 1964 Water Act, created water agencies, required groundwater production over 8 cubic meters per hour (2.22 liters per second) to be declared, and taxed groundwater production (Rinaudo 2020).

The 1935 Water Act primarily concerned the industrial use of groundwater—the primary use of groundwater at the time—and therefore exempted agricultural and municipal uses. However, farmers drilled thousands of new wells after droughts in the 1970s and 1980s with the resulting production drying up many springs, streams, and wetlands (Rinaudo 2020). In 1979, a working group for industry and mines of Ile-de-France recommended prioritizing use of the Albian Aquifer for drinking water and limiting production to 18 million cubic meters per year (Vernoux 2004). By the mid-1990s, water levels at the Grenelle Well had dropped to about 20 meters above sea level (Dupaigne and Violette 2020).

In response to increased production and continuing water-level declines, the French government repealed the 1935 Water Act and passed the 1992 Water Act (Rinaudo 2020). The 1992 Water Act restricted the ability of landowners to produce groundwater and brought agricultural water use under federal regulation, requiring metering of certain water users and creating water restriction areas where existing wells could be forced to reduce production and new wells could be banned (Rinaudo 2020). The Act also required the development of local water management plans created with stakeholders to balance the demands of different users and maintain aquatic environments, including the reduction or cessation of groundwater rights according to critical groundwater levels to guarantee municipal supplies and water for the environment (Rinaudo 2020).

In 2003, the working group for industry and mines of Ile-de-France established that groundwater production would be reduced if water levels in the Albian Aquifer were lower than 31 meters above sea level (Vernoux 2004). By 2018, water levels at the Grenelle Well had increased to about 40 meters above sea level (Dupaigne and Violette 2020).

Although the Albian Aquifer was able to come into equilibrium, including at level that allowed the artesian wells of Paris to flow again, other basins experienced issues with farmers pumping more as the water-level threshold approached (Rinaudo 2020). The Albian Aquifer did not have this problem since most of the production was for drinking water and not agriculture. For those basins with that problem, several managers capped the total volume that could be produced and then allocated production permits to producers (Rinaudo 2020). The 2006 Water Act required the assessment of maximum production volume to meet environmental goals in at least four out of five years, leading to decentralized management of groundwater production (Rinaudo 2020).

Legal Framework: Although there is a foundation under the Rule of Capture, gradual federal regulation culminated with federal restrictions on use.

Actors and Policies: The federal government and various users.

Information and Knowledge: Information and knowledge increased with time culminating with enough of an understanding to regulate the aquifer and access impacts from management.

Why Is This Legislatively Forced Sustainability? The legislature, though a series of actions over several decades, forced the sustainable production of groundwater, returning artesian pressures above the land surface.

Desire Driven: The Barton Springs Segment of the Edwards Aquifer

Flowing at an average of 1400 liters per second, Barton Springs, named after an early settler at the springs, had long been a source of water to Native Americans and hosted three Spanish Missions in the 1700s (Brune 1981). After Texas won independence from Mexico, a committee selected the area to host the new capitol based, in part, on the springs, describing them as "…perhaps the greatest and most convenient water power to be found in the republic" (Brune 1981). Besides providing a supply of water for drinking and operating a mill, Austinites picnicked and swam at the springs. The nearly constant 21 °C temperature is a welcome respite from hot summers. These days, almost 800,000 people visit the springs every year (COA 2022). To many, Barton Springs is considered the soul of Austin (Pipkin 1993).

Barton Springs is fed by the Barton Springs Segment of the Edwards Aquifer, which stretches from the Colorado River, close to where Barton Springs discharges, about 30 kilometers west to the dynamic Kyle groundwater divide near the City of Kyle (Fig. 8.1). The aquifer is about 5 to 15 kilometers wide with a contributing zone extending as much as 60 kilometers west funnels runoff to losing streams that cross the recharge zone.

As the Austin area grew, locals developed concerns about the quality and quantity of water in the aquifer and the springs. After most of the municipalities that relied on the aquifer and landowners prepared a petition in 1985 to create a groundwater conservation district over the aquifer, the Texas Legislature created the Barton Springs Edwards Aquifer Conservation District in 1987 (Gardner 2017; BSEACD 2022).

In 1992, concerned citizens formed the Save Our Springs Coalition to strengthen watershed protection over the aquifer and in the contributing zone as development extended southwestward (SOSA 2022). The Coalition succeeded in getting the Save Our Springs Ordinance passed to protect water quality to the springs and in defending the ordinance in court.

In 1992, the Coalition petitioned the United States Fish and Wildlife Service to list the Barton Springs Salamander (*Eurucea sosorum*, named in honor of the Save Our Springs Ordinance) as endangered, something that required two legal battles before it was listed in May of 1997. Because the species only exists at Barton Springs and a few other localities in the area and relies on springflow for its existence, continued springflow—and thus sustainable management—is required. Therefore, the Alliance sought a listing to provide additional protection to the quality and quantity of water in the aquifer and springs (Giller 2017). Concerns over development and its impact to the springs continued, as shown by a film produced by the actor Robert Redford and the Academy Award-nominated director Terrence Malick (Malick and others 2008).

The groundwater district used its data collection, studies on endangered species, a numerical groundwater flow model (Scanlon and others 2003), and a stakeholder process to identify unacceptable environmental, economic, or social consequences and then calculate a sustainable yield. Ultimately, the sustainable yield was determined by maintaining minimum flows at the springs during a repeat of the drought of record and maintaining saturated thickness in the aquifer in the recharge zone for household (and other) wells. The district honors historical use, although permits are subject to reductions during droughts, and offers interruptible permits that are suspended during times of drought. In developing its regulatory framework, the district also strives to protect the spring habitats of relevant endangered species.

Legal Framework: Legislative action at the state level and regulatory action at the local level. There is also motivation to comply with the federal Endangered Species Act.

Actors and Policies: Concerned municipalities and landowners that initially filed a petition to create a district, the state legislature that created the district, and the district.

Information and Knowledge: Stakeholder knowledge of the cultural importance of the springs was high; federal and state agencies and academics had collected and interpreted data and developed a numerical groundwater flow model; and the district has hired staff and consultants to advance understanding of the resource.

Why Is This Desire-Driven Sustainability? This aquifer is an example of desire-driven sustainability because local activism and action led directly to sustainable management. There was also the perceived threat of federal intervention over endangered species.

Hydrologically Forced and De Facto: The Laugarnes Geothermal System in Reykjavik, Iceland

Legend has it that, after Ingólfur Arnarson approached Iceland in his ship around 870 AD, he threw his high-seat pillars overboard so the Norse gods could guide him to where he should settle (Hjálmarsson 2007). Almost three years later, Arnarson's slaves found the pillars on the southwestern shore of the island, and Arnarson established his permanent home at the discovery site (Hjálmarsson 2007). Arnarson named his homestead Reykjavík (reykja for smokey, vík for bay) after the steam from nearby hot springs (Hjálmarsson 2007).

Iceland has many hot springs and geysers because of its location on the Mid-Atlantic Ridge, an area of active volcanic activity as magma creeps up from the depths and spreads laterally outwards in opposite directions, slowly separating the North American and Eurasian continents in the Northern Atlantic (Fig. 8.11). As a result, Iceland grows wider in the east-west direction by an average of 18.9 to 20.2 millimeters per year (Geirsson and others 2006). Because of the ridge, Iceland is predominately made of basalts and is geologically young with the oldest basalts at 16 million years old (Jónsson and others 2014).

Iceland is almost entirely dependent on groundwater for drinking water, heating, and electricity for municipal purposes. About 95 percent of drinking water is sourced from groundwater (Gunnarsdottir and others 2015; Kløve and others 2017). About 89 percent of household heating is sourced from groundwater-based geothermal sources (Axelsson and others 2010). And about 31 percent of electricity in Iceland is sourced from deeper geothermally heated groundwater with the remainder (69 percent) from hydropower, sourced by streams with a groundwater

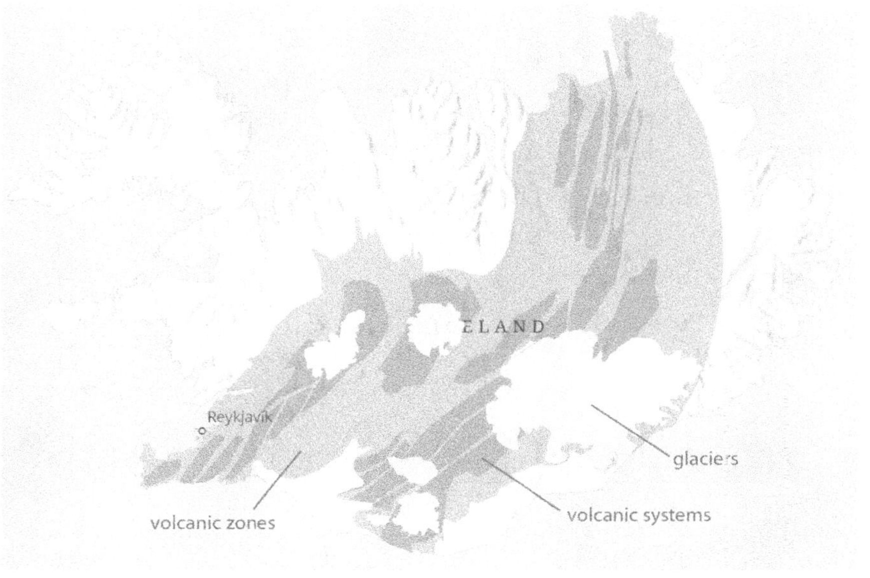

Fig. 8.11 The volcanic zones and systems of Iceland (map generated by the author in ArcGIS and modified in Adobe Illustrator; sources: Esri, Airbus DS, USGS, NGA, NASA, CGIAR, N Robinson, NCEAS, NLS, OS, NMA, Geodatastyrelsen, Rijkwaterstaat, GSA, Geoland, FEMA, and the GIS user community; aquifer data from MetteSvaneDK)

component[11] (Orkustofnun 2021) with much of the hydropower dedicated to the smelting of aluminum. As Sigurdsson (1987) noted, "Groundwater is an energy carrier."

Groundwater systems in Iceland are divided into cold groundwater systems and geothermal systems with geothermal systems further divided into low temperature (<= 150 °C in the upper 1000 meters) and high temperature (> 200 °C at 1000 meters in depth) systems (Franzen and others 1997). Cold groundwater typically has total dissolved solids less than 100 milligrams per liter while low temperature groundwater is about 200 to 400 milligrams per liter and high temperature groundwater is about 1000 to 2000 milligrams per liter (Franzen and others 1997).

[11] Slight amounts of electricity are generated by wind (0.04 percent) and fuel (0.02 percent).

Fig. 8.12 Þvottalaugar hot springs in Reykjavik, Iceland, used for laundry (photo by Hans Petersen, scanned postcard from the author's collection)

Young lava flows can behave like karst systems with rapid recharge and flow and are therefore susceptible to some of the same issues such as being easily contaminated (Kiernan and others 2003). For example, water traveled as fast as 77 meters per day in one lava flow in Iceland (Ketilsson et al. 2017). Besides the basaltic bedrock there are also glacial outwash plains around the island, some of them substantial (tens of kilometers wide and hundreds of meters thick) due to glacial outburst floods called jökulhlaups (Robinson and others 2008).

One of the hot springs in Reykjavík was Þvottalaugar (þvotta for washing, laugar for pools) which were used for washing clothes (Fig. 8.12). The springs are located in what is known today as the Laugarnes Geothermal System and their location can still be visited in the park in Laugardalur, also known as Hot Spring Valley.[12] Laugavegur, a primary retail street in today's downtown Reykjavík, was the pathway from the

[12] Steam from this area is probably why Arnarson named the area Reykjavík.

city center to the spring.[13] Before groundwater production started in the area, the springs flowed about 5 to 10 liters per second (Thornsteinsson and Elíasson 1970).

Between 1928 and 1930, 14 wells were drilled near the springs to tap into hot water for heating in the city (Thornsteinsson and Elíasson 1970). These wells, the deepest at 246 meters, flowed about 15 to 20 liters per second (Thornsteinsson and Elíasson 1970). Two more wells were drilled in the area in 1940 and an additional 16 wells between 1956 and 1959, the latter of which produced a flow of 60 liters per second (Thornsteinsson and Elíasson 1970). An additional 22 wells drilled between 1959 and 1963 each flowed between 1 to 50 liters per second (Thornsteinsson and Elíasson 1970). Starting in 1960, turbine pumps were used to increase production from the wells (Thornsteinsson and Elíasson 1970) with the maximum depth of the wells at 3000 meters (Bodvarsson and Zais 1978). Production is currently about 150 liters per second (Gunnlaugsson and Ívarsson 2010) from 10 wells resulting in the world's largest geothermal heating service (Axelssona and others 2010).

Hot water and power sourced from geothermal groundwater in Iceland are typically open loop systems, which means that groundwater is pumped, used, and then disposed. For example, water for the world-famous Blue Lagoon is effluent from the nearby geothermal power plant.

Thornsteinsson and Elíasson (1970) suggested that the hot springs were affected after the first wells were drilled and produced in the area.[14] After pumps started to be used in 1960, water-levels in the aquifer dropped 120 meters and stopped all artesian flows (Axelssona and others 2010) and, presumably, flows at Þvottalaugar, which are dry today. Pumping increased in the 1980s and resulted in another 30 meters of water-level declines which induced unwelcome inflows of cold fresh water and slightly saline waters (Axelssona and others 2010). The change in geochemistry resulted in calcite-scaling issues in the distribution

[13] As providence would have it, I am writing this section in a hotel on Laugavegur!

[14] "The depth of the deepest well was 246 meters and together they flowed 15-20 liters per second at a temperature of 95°C, as compared to 5-10 liters per second previously issuing from the spring." Interestingly, I was not able to find a reference to the springs drying up in response to groundwater production. Modern-day photos show that it is dry.

system, an issue that abated after production returned to pre-1980 levels in the 1990s (Axelssona and others 2010).

Groundwater production in Iceland has lowered water levels and impacted springs and discharge to streams (Franzen and others 1997). About 46 geysers in Iceland have been quenched from geothermal groundwater production, about 75 percent of all geysers in the country (Barrick 2010). Seawater has infiltrated into some geothermal reservoirs in the coastal areas (Franzen and others 1997).

Septic systems have caused some water-quality problems. A norovirus outbreak at a hotel on Lake Mývatn was caused by a septic tank 80 meters upstream from the hotel's water well (Gunnardottir and others 2013). *E. coli* has been found in well water in Vogarf (Arnadottir 2014) as well as in Reykjavik (Knobloch and others 2020, 2021). In a report on the implementation of the European Union's Water Framework Directive, Jónsson and others (2014) identified a groundwater body "at risk" at the airport/former NATO airbase and four other groundwater bodies as "possibly at risk."

Axelssona and others (2010) concluded that most of the low-temperature geothermal systems being produced were in equilibrium. Jónsson and others (2014) reported no information on "groundwater bodies with directly dependent surface water ecosystems or terrestrial ecosystems, using nationally developed criteria."

Olafsson and others (2014) concluded that economic activities in Iceland "have not occurred in accordance with commonly accepted notions of sustainable development, which is focused upon the reconciliation of economic, environmental and social objectives, together with recognition of current and future need across these domains" and that Iceland considers water quality but not water quantity. Although current geothermal production may be in equilibrium, there are concerns that power will be generated for export to Europe once Iceland is connected to the European grid.

For the most part, Iceland does not regulate groundwater production. However, if a well produces more than 70 liters per second, the well owner is required to receive a permit from the National Energy Authority (Gunnarsdottira and others 2014), suggesting that these permits are for larger geothermal wells.

Given its location near the Arctic Circle, Iceland is particularly susceptible to climate change. Climate change is expected to affect groundwater water levels, groundwater temperatures, and the timing of snowmelt recharge (Kløve and others 2017). Glacier retreat is also expected to impact local groundwater systems (Dochartaigh and others 2019). Microbial contamination of groundwater can (and does) occur during periods of high meltwater and rainwater runoff (Knobloch and others 2020), something that can be expected to increase in the future due to climate change. More dramatically, as glaciers retreat and the land surface rebounds, volcanic eruptions and glacial floods are expected to increase (Ágústsdóttir 2015) with concomitant impacts to groundwater.

Legal Framework: There is no regulatory framework except a requirement to get a permit for larger wells.

Actors and Policies: Private companies, communities, the federal government, academia, and environmental groups.

Information and Knowledge: Not much is known about groundwater resources except from experience at well sites and fields. Iceland is unique in that it produces groundwater for fresh cold water, geothermal heating, and geothermal electric generation. For geothermal uses, users have to worry about not only saltwater encroachment but also cold water encroachment. Since all these waters are meteoric (rainwater sourced), there is the possibility of interference of the different uses, although hot geothermal waters tend to be more distant from warm geothermal waters and cold freshwater.

Why Is This Hydrologically Forced and De Facto Sustainability? Groundwater production in Reykjavík for warm geothermal waters induced seawater intrusions which caused scaling issues. It appears this production was reduced to reduce the impact of scaling, therefore production was hydrologically forced to a lower, sustainable level. This is de facto sustainability because there is no regulatory framework for ensuring the sustainable development of cold, warm, and hot groundwater, except, perhaps, where the production of warm and hot groundwater may impact flows at remaining springs and geysers.

De Facto: Val Verde County, Texas

Val Verde County borders Mexico with the Rio Grande and holds, along with Kinney County directly to the east, what could be called the Southern Segment of the Edwards Aquifer (Fig. 8.1). Two large springs issue from the Edwards Aquifer in the county: San Felipe Springs in Del Rio (average flow of 2600 liters per second), which also serves as the water supply for the city and the Laughlin Air Force Base, and Goodenough Springs (average flow of 3900 liters per second), which now rests below the surface of Lake Amistad, a reservoir on the Rio Grande shared by Mexico and the United States (Brune 1981). Numerous other springs, such as Finegan Springs and Dolan Springs, also flow in the county (Brune 1975). Two spring-fed rivers flow south across the county and feed into the reservoir: the Pecos River and the Devils River.

Much of this book has been about human activity decreasing springflows, but, in the case of San Felipe Springs, springflows have increased, possibly as much as 50 percent, due to the creation and inundation of Lake Amistad, which raised water levels in the aquifer thus increasing flows to the springs. Flow from Goodenough Springs is presumably less due to the water column (41 meters when the conservation pool for the reservoir is full) now resting above it (perhaps assisting in increasing flows to San Felipe Springs).

Val Verde County is relatively distant from nearby cities that may be seeking additional water supplies: 230 kilometers from Laredo, 235 kilometers from San Antonio, 240 kilometers from San Angelo, and 300 kilometers from Midland as well as about 200 kilometers from water-intense hydraulic fracturing in the Eagle Ford Shale and about 300 kilometers from the Permian Basin. Regardless, a large, presumably sustainable supply of groundwater has prompted several landowners to market their groundwater to potential buyers, but distance (and the associated cost of pipelines and pumping) has thus far kept a large groundwater-export project from being realized. As a result, despite no groundwater conservation district and thus no groundwater management, the aquifer is being used sustainably. But for how long?

Transporting groundwater long distances for municipal use is not unheard of in Texas. Besides the aforementioned 225-kilometer pipeline from the Wilcox Aquifer to San Antonio there is the 400-kilometer pipeline conjunctively transporting groundwater and surface water across the High Plains and a proposed 150-kilometer pipeline from the Leon-Belding Aquifer to Midland to transport brackish groundwater for desalination.

Although Val Verde County has considered creating a groundwater conservation district to manage their groundwater resources many times over the past decade, they have still not formed one. The reasons for this are complicated and multifaceted. Broadly, Texans—particularly rural Texans—tend to be politically conservative; therefore, they generally resist new levels of government and the resulting taxes and fees. These Texans generally need to see the perceived threat to their groundwater resources from external forces exceeding the perceived threat to their groundwater resources from local regulation before agreeing to self-regulation.

In many parts of Texas, since 1949, the perceived threat to groundwater resources was the threat of state control of groundwater, but the more recent threat, over the past 24 years, has been groundwater marketers (Mace 2016). Landowners in Val Verde County concerned about groundwater exports and their potential effects on rivers, streams, and springs as well as the groundwater under their property are further weighing the likelihood of a successful export project against the cons of having groundwater regulated. A groundwater conservation district could be formed after an export project is announced, but the district would be in the unenviable position of potentially challenging investment-backed expectations that could be used to seek takings compensation in court. In other words, it would probably be too late to protect local springflows and private property rights if a district is formed after an export agreement is realized.

One source of hesitation is who controls the district. Rural Val Verde County is sparsely populated compared to Del Rio, so there are concerns that the city would control the district to the detriment of rural landowners. Another source of hesitation is concern that the district will be corrupt and used by groundwater marketers to achieve their goals to the detriment of local users. Another point of contention is the City of Del Rio's desire to have its water allocation codified in the enabling legislation of a groundwater conservation district.

Groundwater marketers tend to prefer no regulation because, under the Rule of Capture, they can produce as much as they want regardless of the consequences. However, municipalities often want to see a permit from a groundwater conservation district before investing in a sizable export project. Despite relying completely on springflow for their water supply, the City of Del Rio seems oddly ambivalent about forming a groundwater conservation district, perhaps because they legally produce surface water (just as the springs erupt to the land surface) rather than groundwater. As populations grow (the outskirts of San Antonio are now only 160 kilometers from the eastern part of the aquifer feeding San Felipe Springs), and, as the climate becomes warmer, more and more communities along the water-starved Interstate-35 growth corridor will be tempted by water resources like the one that provides Del Rio.

Local advocates led by the Devils River Conservancy have advocated for science-driven and watershed-based water management. Local legislators have filed several bills to create a groundwater conservation district in the county over the past decade, but none have passed out of committee let alone a chamber. Local legislators succeeded in getting state agencies to prepare a technical review of the area (Weinberg and others 2018), and the Devils River Conservancy has supported the development of a river basin-scale numerical groundwater flow model (Toll and others 2017). In addition, the Conservancy has worked closely with stakeholders and state agencies to fill data gaps identified by Weinberg and others (2018).

Legal Framework: At present, no regulatory framework under the Rule of Capture.

Actors and Policies: Landowners, water marketers, the City of Del Rio, nongovernmental organizations such as the Devils River Conservancy and the Nature Conservancy, and state legislators. No policies at present but a continuing discussion on whether or not to regulate use and at what level.

Information and Knowledge: Historical information and studies conducted to assess impacts to springflows and review existing information to make policy decisions.

Why Is This De Facto Sustainability? This aquifer is an example of de facto sustainability because the aquifer is currently being produced sustainably through no active management or regulation although the sustainable production may change due to more production. Some locals are trying to form a local groundwater agency to ensure the long-term sustainability of the system (which would change the aquifer to a desire-driven sustainability example).

Strategies for Sustainability

Based on the water budget for an aquifer, there are only so many ways one can achieve sustainability: increase recharge, decrease or optimize production, and manage storage. Groundwater regulators can also manage certain unacceptable environmental, economic, or social consequences.

Increasing recharge to an aquifer depends on the availability of additional water, the ability of the aquifer to accept that additional water (both physically and chemically), and the residence of time of the artificially recharged water. These approaches are generally covered under the specialty of managed artificial recharge (see Fox 2007; Pyne 2019; Dillon and others 2021; Joji and others 2021) and can be achieved through spreading basins, injections wells, and land management. Sources may include surface water, treated wastewater, or other groundwater.

Decreasing production means reducing the production of groundwater whether through water conservation, finding alternative supplies, or buying out users. Optimizing production includes optimizing the placement of wells to maximize allowable production without creating unacceptable consequences. Optimizing may also include managing the timing of production to avoid unacceptable consequences. Permits can be optimized through storing groundwater elsewhere for later use, something that San Antonio does with some of its Edwards groundwater where they store it in a nearby sandstone aquifer for use in times of drought.

Managing storage involves allowing the volume of water stored in the aquifer to increase during non-drought or off-season times and decrease during drought or in-season times (with season referring to the seasonal nature of irrigation). Another example is using excess storage capacity to store water from another source for later retrieval.

It may be possible to manage unacceptable environmental, economic, or social consequences such that more groundwater can be produced. For example, if draining shallow wells is an unacceptable consequence, groundwater managers can have mitigation programs to drill deeper wells for those users (or connect them to an alternative supply). An example of environmental mitigation is the use of injection wells to prevent the intrusion of seawater such as in the Los Angeles area (Jasechko and others 2020). Social consequences may be able to be mitigated with alternative supplies or appropriate financial offsets.

We'll Know the Worth of Water Before the Well Runs Dry

If there's a common thread for achieving sustainable groundwater production or seeking to preserve it, it's people. There was, is, or needs to be somebody with the vision, organizational skills, contacts, and power to promote sustainable groundwater production. These somebodies understood or understand the importance of sustainable management and sought—or are seeking—to institute and implement it. Science often serves the role of providing key information on what is happening and could happen, arming the somebodies with data and analysis to make arguments for sustainable management. In this way, people understand what is at stake before springs, rivers, and wells run dry from nonsustainable production.

Groundwater sustainability may not be for everyone. If a government has to choose between groundwater sustainability and their people dying of thirst, the aquifer is probably going to come up short. In cases where production has far exceeded sustainable levels, such as in the Ogallala Aquifer, it may not be (and has not been) politically possible to reduce production through regulation. But hopefully, at the very least, decision-makers and users in these situations fully understand the short- and long-term consequences of their actions.

Questions

What type of groundwater sustainability exists in your area?
Who are the groundwater actors in your area? Which ones support sustainability and why? Which ones do not support sustainability and why?
Do people in your area understand groundwater issues? Why or why not?

References

Ágústsdóttir, A.M., 2015, Ecosystem approach for natural hazard mitigation of volcanic tephra in Iceland—Building resilience and sustainability: Natural Hazards, v. 78, p. 1669–1691.

Allan, P., 1906, The drought antidote for the North-West: Or, the utilisation of the Artesian Resources of New South Wales: Sydney, Australia, Government Printer.

Angula, H.K., 2001, Foreword *in* Christelis, G., and Struckmeier, W., Groundwater in Namibia—An explanation to the hydrogeological map: Ministry of Agriculture, Water, and Rural Development, Namibia, p 2.

ARMCANZ (Agriculture and Resource Management Council of Australia and New Zealand), 1996, Allocation and use of groundwater—A national framework for improved groundwater management in Australia: Occasional Paper No. 2, Commonwealth of Australia.

Arnadóttir, N., 2014, E.Coli found in drinking water: Grapevine, published September 9, 2014.

AWIC (Artesian Water Investigation Committee), 1945, Artesian water supplies, First Interim Report: Queensland Artesian Water Investigation Committee, Queensland, 40 p.

AWIC (Artesian Water Investigation Committee), 1954, Artesian water supplies in Queensland: Queensland Artesian Water Investigation Committee, Department of the Co-ordinator-General of Public Works, Queensland, 79 p.

Axelssona, G., Gunnlaugsson, E., Jónasson, T., and Ólafsson, M., 2010, Low-temperature geothermal utilization in Iceland—Decades of experience: Geothermics, v. 39, p. 329–338.

Barnett, S., Simmons, C.T., and Nelson, R., 2021, Groundwater resources in Australia—their occurrence, management, and future challenges *in* Mukherjee, A., Scanlon, B.R., Aureli, A., Langan, S., Guo, H., and McKenzie, A.A., editors, Global Groundwater—Source, Scarcity, Sustainability, Security, and Solutions: Chapter 3, Elsevier, Amsterdam, Netherlands, p. 35–46.

Barraqué, B., and de Marsily, G., 2010, Country report for France *in* Groundwater in the Southern Member States of the European Union—An assessment of current knowledge and future prospects: European Academies Science Advisory Council, EASAC policy report 12, SBN: 978-3-8047-2827-1, German Academy of Sciences, 44 p.

Barrick, K.A., 2010, Environmental review of geyser basins—Resources, scarcity, threats, and benefits: Environment Reviews, v. 18, p. 209–238.

Blake, T., and Cook, M., 2006, Great Artesian Basin historical overview: Queensland Department of Natural Resources Mines and Water, Brisbane, Australia

Blumberg, H., and Collins, G., 2016, Implementing three-dimensional groundwater management in a Texas groundwater conservation district: Texas Water Journal, v. 7, no. 1, p. 69–81.

Blumm, M.C., and Ritchie, L., 2005, The pioneer spirit and the public trust—The American Rule of Capture and state ownership of wildlife: Environmental Law, v. 35, n. 4, p. 101–147. https://ssrn.com/abstract=845044.

Bodvarsson, G., and Zais, E., 1978, A field example of free surface testing: in Kruger, P., and Ramey, Jr., H.J., editors, Proceedings Forth Workshop Geothermal Reservoir Engineering, SGP-TR-30, CONF-781222-20, Stanford, California, p. 153–159.

Bollig, M., 2013, Social-ecological change and institutional development in a pastoral community in North-Western Namibia *in* Boling, M., Wotzka, H.P., and Schnegg, M., editors, Pastoralism in Africa—Past, present and futures: London, Berghahn, p. 316–340.

Brake, L, 2020, Development, management and rehabilitation of water bores in the Great Artesian Basin, 1878–2020: Proceedings of The Royal Society of Queensland, v 126, p 153–175.

Brune, G., 1975, Major and historical springs of Texas: Texas Water Development Board Report 189, 94 p. + plates.

Brune, G., 1981, Springs of Texas, Volume 1: Branch-Smith, Inc., Fort Worth Texas, 566 p.

BSEACD (Barton Springs Edwards Aquifer Conservation District), 2022, History: accessed January 4, 2022; https://bseacd.org/about-us/history/.

Bumgarner, J.R., Stanton, G.P., Teeple, A.P., Thomas, J.V., Houston, N.A., Payne, J.D., and Musgrove, M., 2012, A Conceptual Model of the Hydrogeologic Framework, Geochemistry, and Groundwater-Flow System of the Edwards-Trinity and Related Aquifers in the Pecos County Region, Texas: Scientific Investigations Report 2012–5124 (revised July 10, 2012), U.S. Geological Survey, 74 p.

CDWR (California Department of Water Resources), 2020a, California's groundwater and the Sustainable Groundwater Management Act: January 2020, accessed on December 27, 2021: https://water.ca.gov/-/media/DWR-Website/Web-Pages/Programs/Groundwater-Management/Sustainable-Groundwater-Management/Files/CA-Groundwater-and-SGMA-Fact-Sheet.pdf.

CDWR (California Department of Water Resources), 2020b, Sustainable Groundwater Management Act 2019 basin prioritization—Process and results: May 2020, variously paginated (99 total pages).

CDWR (California Department of Water Resources), 2021a, Groundwater: accessed on December 5, 2021, https://water.ca.gov/water-basics/groundwater.

CDWR (California Department of Water Resources), 2021b, Sustainable Groundwater Management Act—Plan evaluation pathways: May 2021, accessed on December 27, 2021, https://water.ca.gov/-/media/DWR-Website/Web-Pages/Programs/Groundwater-Management/Sustainable-Groundwater-Management/Groundwater-Sustainability-Plans/Files/GSP/SGMA-Evaluation-Pathways-Factsheet.pdf.

Christelis, G., Heyns, P., and Struckmeier, W., 2001, Essentials of groundwater *in* Christelis, G., and Struckmeier, W., Groundwater in Namibia—An explanation to the hydrogeological map: Ministry of Agriculture, Water, and Rural Development, Namibia, p 30–39.

Christelis, G., and Struckmeier, W., 2001, Groundwater in Namibia—An explanation to the hydrogeological map: Ministry of Agriculture, Water, and Rural Development, Namibia, 128 p.

CNRA, CEPA, and CDFA (California Natural Resources Agency, California Environmental Protection Agency, and California Department of Food & Agriculture), 2013, California agencies release draft action plan for water, ask for input and dialogue—California Water Action Plan provides roadmap for state efforts: press release, October 31. https://www.waterboards.ca.gov/press_room/press_releases/2013.shtml.

CNRA, CEPA, and CDFA (California Natural Resources Agency, California Environmental Protection Agency, and California Department of Food & Agriculture), 2014, California water action plan: joint publication, 19 p., https://resources.ca.gov/CNRALegacyFiles/docs/california_water_action_plan/2014_California_Water_Action_Plan.pdf.

COA (City of Austin), 2022, Barton Springs Pool: accessed on January 4, 2022; https://www.austintexas.gov/department/barton-springs-pool#:~:text=Depths%20of%20the%20pool%20range,Edwards%20Aquifer%20which%20feeds%20it.

CoAG (Council of Australian Governments), 1994. Communique of February Meeting. Attachment A. National Competition Council, http://ncp.ncc.gov.au/docs/PIAg-001.pdf.

Constitute, 2021, Namibia's Constitution of 1990 with Amendments through 2010: constitureproject.com, 68 p., accessed April 20, 2022.

Contoux, C., Violette, S., Vivona, R., Goblet, P., and Patriarche, D., 2013, How basin model results enable the study of multi-layer aquifer response to pumping: the Paris Basin, France: Hydrogeology Journal, v. 21, p. 545–557.

Cox, R., and Barron, A., editors, 1998, Great Artesian Basin resource study: Great Artesian Basin Consultative Council, 235 p.

Cox, W.G., 1895, Artesian wells as a means of supply: Sapsford & Company, Brisbane, Australia, 146 p.

Dafny, E., 2016, The Great Artesian Basin—Is it that great? Hydrogeology Journal, https://doi.org/10.1007/s10040-016-1444-5.

DAWE (Department of Agriculture, Water and the Environment), 2020, Great Artesian Basin Strategic Management Plan: Department of Agriculture, Water and the Environment, Australian Government, 53 p.

DEWHA (Department of the Environment, Water, Heritage and the Arts), 2009, Water Matters: Australian Government, Issue 4, May, 1 p.

Dickinson, R., Pavley, F., and Rendon, A., 2014, Groundwater management: Assembly Bill No. 1739, https://leginfo.legislature.ca.gov/faces/billTextClient.xhtml?bill_id=201320140AB1739.

Dillon, P., Escalante, E.F., Megdal, S.B., and Massmann, G., editors, 2021, Managed aquifer recharge for water resilience: Water, Special Issue, 408 p. ISBN 978-3-03943-043-7.

Dochartaigh, B.É.Ó., MacDonald, A.M., Black, A.R., Everest, J., Wilson, P., Darling, W.G., Jones, L., and Raines, M., 2019, Groundwater–glacier meltwater interaction in proglacial aquifers: Hydrology and Earth System Sciences, v. 23, no. 11, p. 4527–4539, https://doi.org/10.5194/hess-23-4527-2019

DoE (Department of the Environment), 2014, Great Artesian Basin sustainability initiative (GABSI) value for money review: Department of the Environment, Australia, 52 p.

Dupaigne, T., and, Violette, S., 2020, Impact de l'activité anthropique et des règles de gestion des eaux souterraines sur la ressource – exemple de l'Albien en Île-de-France: La Houille Blanche, 106:1, 16–21, https://doi.org/10.1051/lhb/2020003.

EUWD (Edwards Underground Water District), 1989, Edwards Underground Water District created to conserve, protect, and preserve the Edwards Aquifer: Edwards Underground Water District, 22 p.

Fairfax R.J., and Fensham, R.J., 2002, In the footsteps of J. Alfred Griffiths—A cataclysmic history of Great Artesian Basin springs in Queensland: Australian Geographical Studies, v 40, n 2, p 210–230.

Falk, T., Bock, B., and Kirk, M., 2009, Polycentrism and poverty—Experiences of rural water supply reform in Namibia: Water Alternatives, v 2, p 115–137.

Fensham, R., and Fairfax, R., 2005, Great Artesian Basin water resource plan ecological assessment of GAB springs in Queensland: Department of Natural Resources and Mines, Brisbane, Australia, 27 p.

Fisher, N.H., and Hahn, G.W., 1963, Review of available groundwater data on the Great Artesian Basin: Bureau of Mineral Resources, Geology and Geophysics. Records 1963/128, variously paginated (20 total pages).

Fox, P., editor, 2007, Management of aquifer recharge for sustainability: Acacia Publishing, 641 p.

Franzson, H., Tómasson, J., Sverrisdóttir, and Sigurosson, F., 1997, Geothermal energy in Iceland: *in* Mineral and Thermal Groundwater Resources. Springer, Dordrecht, Chapter 9. https://doi.org/10.1007/978-94-011-5846-6_9.

Frontier Economics, 2016, Economic output of groundwater dependent sectors in the Great Artesian Basin: A report commissioned by the Australian Government and Great Artesian Basin Jurisdictions based on advice from the Great Artesian Basin Coordinating Committee, Melbourne, Australia, 60 p.

GABCC (Great Artesian Basin Consultative Council), 2000, Great Artesian Basin Strategic Management Plan: Great Artesian Basin Consultative Council for the Australian Government Department of Agriculture, Fisheries and Forestry, Canberra, 60 p.

Gardner, M., 2017, Happy BD to BSEACD! Conservation district celebrates 30 years: Hays Free Press, p. 1C-2C.

Gaye, C.B., and Tindimugaya, C., 2019, Review—Challenges and opportunities for sustainable groundwater management in Africa: Hydrogeology Journal, v 27, p 1099–1110. https://doi.org/10.1007/s10040-018-1892-1.

Geirsson, H., Árnadóttir, T., Völksen, C., Jiang, W., Sturkell, E., Villemin, T., Einarsson, P., Sigmundsson, F., and Stefánsson, R., 2006, Current plate movements across the Mid-Atlantic Ridge determined from 5 years of continuous GPS measurements in Iceland: Journal of Geophysical Research—Solid Earth, https://doi.org/10.1029/2005JB003717.

Giller, G., 2017, A fragile balance—The endangered Barton Springs salamanders of Texas: Reptiles, accessed January 4, 2022; https://reptilesmagazine.com/a-fragile-balance-the-endangered-barton-springs-salamanders-of-texas/.

Griggs, B.W., 2021, Reaching consensus about conservation: High Plains lessons for California's Sustainable Groundwater Management Act: University of the Pacific Law Review, v. 52, n. 3, p. 495–547.

Gulley, R.L., 2015, Heads above water—The inside story of the Edwards Aquifer Recovery Implementation Program: Texas A&M University Press, 234 p.

Gunnarsdottir, M.J., Gardarsson, S.M., and Andradottir, H.O, 2013, Microbial contamination in groundwater supply in a cold climate and coarse soil—Case study of norovirus outbreak at Lake Mývatn, Iceland: Hydrology Research, 44.6, p. 1114–1128.

Gunnarsdottira, M.J., Gardarsson, S.M., and Bartram, J., 2014, Developing a national framework for safe drinking water—Case study from Iceland: International Journal of Hygiene and Environmental Health, IJHEH-12804. https://doi.org/10.1016/j.ijheh.2014.10.003.

Gunnarsdottir, M.J., Gardarsson, S.M., Jonsson, G.S., Armannsson, H., and Bartram, J., 2015, Natural background levels for chemicals in Icelandic aquifers: Hydrology Research, 46.4, p. 647–660.

Gunnlaugsson, E., and Ívarsson, G., 2010, Direct use of geothermal water for district heating in Reykjavík and other towns and communities in SW-Iceland: Proceedings of the World Geothermal Congress 2010, Bali, Indonesia, 6 p.

Habermehl, R.A., 1980, The Great Artesian Basin, Australia: BMR Journal of Australian Geology &: Geophysics, v 5, p 9–38.

Habermehl, R.A., 2018, Groundwater governance in the Great Artesian Basin, Australia *in* Villholth, K.G., López-Gunn, E., Conti, K.I., Garrido, A., and van der Gun, J., Advances in Groundwater Governance: CRC Press, Taylor & Francis Group, p 411–441.

Harris, C., 2002, Culture and geography: South Australia's mound springs as trade and communication routes: Historic Environment, v 16, n 2, p 8–11.
Heyns, P., Seely, M., and Montgomery, S., 1998, Namibia's water—A decision maker's guide: Ministry of Agriculture, Water and Rural Development, Namibia, 173 p.
Heyns, P., and Struckmeier, W., 2001, Introduction—Water in Namibia and rationale for a hydrogeological map project *in* Christelis, G., and Struckmeier, W., Groundwater in Namibia—An explanation to the hydrogeological map: Ministry of Agriculture, Water, and Rural Development, Namibia, p 9–12.
Hinz, M.O., 2016, Customary law ascertained—The customary law of the Nama, Ovaherero, Ovambanderu, and San Communities of Namibia: Volume 3, University of Namibia Press, Windhoek, 570 p.
Hjálmarsson, J.R., 2007, History of Iceland—From the Settlement to the Present Day: Forlagid, Reykjavik, 208 p.
Hovorka, S.D., Mace, R.E., and Collins, E.W., 1998, Permeability structure of the Edwards Aquifer, South Texas—Implications for aquifer management: Bureau of Economic Geology, The University of Texas at Austin, Report of Investigations No. 250, 55 p.
Jasechko, S., Perrone, D., Seybold, H., Fan, Y., and Kirchner, J.W., 2020, Groundwater level observations in 250,000 coastal US wells reveal scope of potential seawater intrusion: Nature Communications, 11:3229 | https://doi.org/10.1038/s41467-020-17038-2.
Joji, V.S., Viswanathan, S., and Sebastian, S., 2021, A textbook of managed aquifer recharge: Scholars' Press, KS Omniscriptum Publishing, 120 p.
Jokisch, A., Schulz, O., and Papangelou, A., 2018, Geographic conditions of central-northern Namibia *in* Liehr, S., Kramm, J., Jokisch, A., and Müller, K., editors, Integrated Water Resources Management in Water-Scarce Regions—Water Harvesting, Groundwater Desalination and Water Reuse in Namibia: IWA Publishing, p 10–18.
Jónsson, G.S., Jensson, H., Weisshappel, J.B., Þórðarsson, T., and Karlsdóttir, S.D., 2014, Water Framework Directive (2000/60/EC), Article 5 (and 6), Summary Report—Characteristics and impact analyses: Environment Agency of Iceland, 29 p.
Ketilsson, J., Óskarsdóttir, S.M., Claesson, A., and Collett, N.J., 2017, The Kerauga Cave and Lækjarbotnaveita in South Iceland—Groundwater safety and hydrogeology: Journal of Water Security, v. 3, https://doi.org/10.15544/jws.2017.001.
Kiernan, K., Wood, C., & Middleton, G., 2003. Aquifer structure and contamination risk in lava flows–Insights from Iceland and Australia: Environmental

Geology. v. 43, no. 7, p. 852–865, https://doi.org/10.1007/s00254-002-0707-8

Kløve, B., Margrethe, H., Kvitsand, L., Pitkänen, T., Gunnarsdottir, T.J., Gaut, S., Gardarsson, S.M., Rossi, P.M., and Miettinen, I., 2017, Overview of groundwater sources and water-supply systems, and associated microbial pollution, in Finland, Norway and Iceland: Hydrogeology Journal, v. 25, p. 1033–1044.

Kluge, T., and Polak, M., 2018, Water supply infrastructure and sanitation in central-northern Namibia *in* Liehr, S., Kramm, J., Jokisch, A., and Müller, K., editors, Integrated Water Resources Management in Water-Scarce Regions—Water Harvesting, Groundwater Desalination and Water Reuse in Namibia: IWA Publishing, p 19–22.

Knobloch, S., Klonowski, A., and Marteinsson, V.Þ, 2020, Microbial diversity of the Heiðmörk groundwater system and potential sources of drinking water contamination: Skýrsla Matís 06–20, January, 40 p.

Knobloch, S., Klonowski, A., Tómasdóttir, S., Kristjánsson, B.R., Guðmundsson, S., and Marteinsson, V.Þ, 2021, Microbial intrusion and seasonal dynamics in the groundwater microbiome of a porous basaltic rock aquifer used as municipal water reservoir: FEMS Microbiology Ecology, v. 97, no. 3, https://doi.org/10.1093/femsec/fiab014.

Lawrence, C.R., 2018, A history of hydrogeology in Australia from pre-European to the 21st Century: Modern Environmental Science and Engineering, v. 4, n. 4, p. 276–308.

Leahy, T.C., 2016, Desperate times call for sensible measures—The making of the California Sustainable Groundwater Management Act: Golden Gate University School of Law Environmental Law Journal, v. 9, n. 1, p. 5–40.

Mace, R.E., 2016, So secret, occult, and concealed—An overview of groundwater management in Texas: in Law of the Rio Grande, Water Law Institute, CLE International, April 14–15, 2016, La Fonda, Santa Fe, New Mexico, Section H3, 27 p.

Mace, R.E., 2021a, Five gallons in a ten-gallon hat—Groundwater sustainability in Texas: The Meadows Center for Water and the Environment Report 2021-08, Texas State University, 50 p.

Mace, R.E., 2021b, Stakeholder engagement in water resources decisions in Texas: American Association of Geographers Annual Meeting, April 7–11, 2021.

Mace, R.E., Leurig, S., Seely, H., and Wierman, D.A., 2020, Bringing back Comanche Springs—An analysis of the history, hydrogeology, policy, and economics: Report 2020-08, The Meadows Center for Water and the Environment, Texas State University, 153 p.

Macdonald, A.M., Calow, R.C., Macdonald, D.M.J., Darling, W.G., and Dochartaigh, B.É.Ó., 2009, What impact will climate change have on rural groundwater supplies in Africa? Hydrological Sciences–Journal–des Sciences Hydrologiques, v 54, n 4, p 690–703.

Malick, T., Redford, R., Dunn, L., Sewell, D., Warren, W., Sewell, J., and Daniel, L., 2008, The Unforeseen: Two Birds Films, 92 minutes.

Mapani, B.S., 2005, Groundwater and urbanisation, risks and mitigation—The case for the city of Windhoek, Namibia: Physics and Chemistry of the Earth, v 30, p 706–711.

Maréchal, J.-C., and Rouillard, J., 2020, Groundwater in France—Resources, use, and management Issues—Sustainable Groundwater Management A Comparative Analysis of French and Australian Policies and Implications to Other Countries, p.17–45, https://doi.org/10.1007/978-3-030-32766-8_2. hal-02512154.

MAWRD (Ministry of Agriculture, Water and Rural Development), 2000, National water policy white paper: Republic of Namibia, 45 p.

McKay, J., 2002, Encountering the South Australian landscape—Early European misconceptions and our present water problems: Hawke Institute Working Papers Series, No. 21, Hawke Institute, University of South Australia, Magill, South Australia, 36 p.

Meunier, M., 1978, Les puits artésiens de l'Avallonnais: Académie du Morvan, v. 5, n. 6, p. 3–14. https://gallica.bnf.fr/ark:/12148/bpt6k96033964/f5.item.r=grenelle%20puit.

Milman, A., and Kiparsky, M., 2020, Concurrent governance processes of California's Sustainable Groundwater Management Act: Society & Natural Resources, v. 33, n. 12, p. 1555–1566.

Mudd, G.M., 2000, Mound springs of the Great Artesian Basin in South Australia—A case study from Olympic Dam: Environmental Geology, v 39, n 5, p 463–476.

Nicol, S., 2005, The Great Artesian Basin: Department of Natural Resources and Water, The State of Queensland, accessed April 10, 2022, https://web.archive.org/web/20070315203118/http://www.nrw.qld.gov.au/water/gab/.

Nhuleipo, O., 2022, Namibia: Gaborone Declaration for Sustainability in Africa, accessed April 24, 2022. http://www.gaboronedeclaration.com/namibia.

NRMMC (Natural Resource Management Ministerial Council), 2002, Managing overallocated groundwater systems: Commonwealth of Australia.

Olafsson, S., Cook, D., Davidsdottir, B., & Johannsdottir, L., 2014, Measuring countries' environmental sustainability performance – A review and case study of Iceland: Renewable and Sustainable Energy Reviews 39934–948 S136403211400553X 10.1016/j.rser.2014.07.101

Orkustofnun, 2021, Orkustofnun Data Repository OS-2021-T009-01.

Parkinson, C.E., 1954, Minority Report *in* Artesian Water Investigation Committee, Artesian water supplies in Queensland: Queensland Artesian Water Investigation Committee, Department of the Co-ordinator-General of Public Works, Queensland, p 70–79.

Paterson, A.B., 1902, Rio Grande's last race and other verses: Angus and Robertson, Melbourne.

Pavley, F., Dickinson, R., and Rendon, A., 2014, Sustainable groundwater management: Senate Bill No. 1168, http://www.leginfo.ca.gov/pub/13-14/bill/sen/sb_1151-1200/sb_1168_bill_20140822_amended_asm_v92.htm

Pipkin, T., 1993, Barton Springs—The soul of a city: Softshoe Pub, 135 p.

Pittman, E.F., and David, T.W.E., 1903, Irrigation geologically considered with special reference to the artesian area of New South Wales: Journal and Proceedings of the Royal Society of NSW, v 37, p CIII–CLIII.

Polak, M., 2014, Zwischen Haushalt und Staat—Lokale Water Governance im zentralen Norden Namibias (Between Household and State—Local Water Governance in Central Northern Namibia): Forum Politische Geographie, Book 12, LIT-Verlag Münster, Berlin, 240 p.

Polak, M., and Kluge, T., 2018, Water governance in Namibia—Progress in and challenges of putting IWRM into practice *in* Liehr, S., Kramm, J., Jokisch, A., and Müller, K., editors, Integrated Water Resources Management in Water-Scarce Regions—Water Harvesting, Groundwater Desalination and Water Reuse in Namibia: IWA Publishing, p 23–26.

Powell, O., Silcock, J., and Fensham, R., 2015, Oases to oblivion—The rapid demise of springs in the South-Eastern Great Artesian Basin, Australia: Groundwater, v 53, n 1, p. 171–178.

Pyne, D.G., 2019, Groundwater recharge and wells—A guide to aquifer storage recovery: CRC Press, 400 p.

Ravenstein, E.G., 1900, The voyages of Diogo Cão and Bartholomeu Dias, 1482–88: The Geographic Journal, variously paginated.

Remmert, D., 2016, Water governance in Namibia—A tale of delayed implementation, policy shortfalls, and miscommunication: Special Briefing Report No. 13, Democracy Report, 20 p.

Rinaudo, J.-D., Holley, C., Barnett, S., and Montginoul, M., 2020, Sustainable Groundwater Management—A Comparative Analysis of French and Australian Policies and Implications to Other Countries: Springer, 562 p.

Rinaudo, J.-P., 2020, Groundwater policy in France: From private to collective management in Rinaudo, J.-P., Holley, C., Barnett, S., and Montginoul, M., editors, Sustainable Groundwater Management—A Comparative Analysis of French and Australian Policies and Implications to Other Countries: Chapter 3, Global Issues in Water Policy, Volume 24, Springer, p. 47–65.

Robinson, Z.P., Fairchild, I.J., and Russell, A.J., 2008, Hydrogeological implications of glacial landscape evolution at Skeiðarársandur, SE Iceland: Geomorphology, v. 97, p. 218–236.

Rudestam, K., and Langridge, R., 2014, Sustainable yield in theory and practice—Bridging scientific and mainstream vernacular: Groundwater, v. 52, p. 90–99.

Sax, J.L., 2002, We don't do groundwater—A morsel of California legal history: University of Denver Water Law Review 6:269.

Scanlon, B.R., Mace, R.E., Barrett, M.E., and Smith, B., 2003, Can we simulate regional groundwater flow in a karst system using equivalent porous media models? Case study, Barton Springs Edwards aquifer, USA: Journal of Hydrology, v. 276, n. 1, p. 137–158. https://doi.org/10.1016/S0022-1694(03)00064-7.

Schnegg, M., Bollig, M., and Linke, T., 2016, Moral equality and success of common-pool water governance in Namibia: Ambio, v 45, p 581–590. https://doi.org/10.1007/s13280-016-0766-9.

Schneider, G.I.C., Schneider, M.B., and du Pisani, A.L, 2001, Natural and Socio-Economic Environment *in* Christelis, G., and Struckmeier, W., Groundwater in Namibia—An explanation to the hydrogeological map: Ministry of Agriculture, Water, and Rural Development, Namibia, p 14–29.

Sigurdsson, F., 1987, Hydrogeology and geohydrology: Report 6, National Energy Authority, Reykjavik, Iceland, 28 p. + 21 plates.

Smith, M., Cross, K., Paden, M. and Laban, P., editors, 2016, Spring—Managing groundwater sustainably: International Union for Conservation of Nature, Gland, Switzerland, 133 p.

Soetaert, F., Wanke, H., Dupuy, A., Lusuekikio, V., Gaucher, E.C., Bordmann, V., Fleury, J.-M., and Franceschi, M., 2022, Toward the sustainable use of groundwater springs—A case study from Namibia: Sustainability, v 14, 3995. https://doi.org/10.3390/su14073995.

Solanes, M., and Gonzalez-Villarreal, F., 1999, The Dublin Principles for water as reflected in a comparative assessment of institutional and legal arrangements for Integrated Water Resources Management: TAC Background Papers No. 3, Technical Advisory Committee, Global Water Partnership, 43 p.

SOSA (Save Our Springs Alliance), 2022, The story of the Save Our Springs Alliance: accessed January 4, 2022; https://www.sosalliance.org/our-story.html.

SWRCB (State Water Resources Control Board), 2013, Groundwater Workplan Concept Paper—Discussion Draft: October 4th, http://www.waterboards.ca.gov/water_issues/programs/groundwater/docs/gw_workplan100713.pdf.

Thompson, Jr., B.H, M. Rohde, Howard, J.K., and Matsumoto, S., 2021, Mind the gaps—The case for truly comprehensive sustainable groundwater management: Water in the West. Stanford Digital Repository. Available at: https://purl.stanford.edu/hs475mt1364.

Thornsteinsson, T., and Elíasson, J., 1970, Geohydrology of the Laugarnes Hydrothermal System in Reykjavík, Iceland: National Energy Authority of Iceland, March 1970, 29 p. + 11 figures.

Toll, N., Fratesi, S.B., Green, R.T., Bertetti, F.P., and Nunu, R., 2017, Water resource management of the Devils River watershed, final report: Southwest Research Institute report under contract to the Devils River Conservancy, at http://www.devilsriverconservancy.org/research/, accessed January 2022.

Turral, H., and Fullagar, I., 2007, Institutional Directions in Groundwater Management in Australia *in* Giordano, M., and Villholth, K.G., editors, The agricultural groundwater revolution—Opportunities and threats to development: CAB International, p. 320–361.

Umenne, S., and Chirwa, V., 2012, Culture and environment as generators of sustainable design—The case of Hoachanas Village, Hardap Region, Namibia: Sustainable Futures: Architecture and Urbanism in the Global South Kampala, Uganda, June 27–30, p 167–176.

United Nations, 1992, Agenda 21: United Nations Conference on Environment & Development, Rio de Janerio, Brazil, 3 to 14 June 1992, 351 p.

USFWS (United States Fish & Wildlife Service), 2021, Endangered Species Act—A history of the Endangered Species Act of 1973: accessed December 28, 2021: https://www.fws.gov/endangered/laws-policies/esa-history.html.

van Rensburg, P., 2018, Mitigation through innovation—Sustainable water supply in Windhoek: Urbanet, posted March 22, 2018; accessed April 24, 2022. https://www.urbanet.info/mitigation-through-innovation/.

Vernoux,·J.F., 2004, Réseau piézométrique du bassin Seine-Normandîe, Stations de surveillance de ta nappe de l'Albîen: BRGM, Report BRGM/RP-53009-FR, variously paginated (33 total pages).

Votteler, T.V., 1998, The little fish that roared—The Endangered Species Act, state groundwater law, and private property rights collide over the Texas Edwards Aquifer: Environmental Law, v. 28, n. 4, p. 845–879.

Wallace, M., 2011, A history of Namibia—From the beginning to 1990: Columbia University Press, New York, 451 p.

WCED (World Commission on Environment and Development), 1987, Our Common Future: Oxford University Press, 400 p.

Weinberg, A., French, L.N., and Perez, J.B., 2018, Overview of groundwater conditions in Val Verde County, Texas: Texas Water Development Board, 227 p.

Wilkinson, C.S., 1882, Notes on the occurrence of artesian wells in the Albert District, New South Wales: Proceedings of the Linnean Society of NSW, n 6, p 155–157.

Williamson, D.H., 1996, The development of hydrogeology in Australia: self-published book, Ermington, N.S.W., ISBN 0646411683, 50 p.

Williamson, W.H., 2012, The history of hydrogeology in Australia *in* Howden, N., and Mather, J., editors, History of Hydrogeology, CRC Press, p 7–27.

Woolley, D.R., Habermehl, M.A., Palmer, J., Sibenaler, X., and O'Neill, D., 1987, Improved Efficiency of Artesian Groundwater Use, Rehabilitation of Bores and Conversion from Drains to Piping: Report to the Australian Water Resources Council Groundwater Committee by the Great Artesian Basin Sub-Committee, 18 p.

9

Groundwater Will Be Produced Sustainably Whether We Like It or Not

"[humans] cannot escape the consequences of their choices about which Nature will have the last word."
Jonas Salk (*1992*)

As the artesian wells stopped flowing in the Trinity Aquifer along the Blackland Prairies of Texas, dreams of a limitless supply of groundwater ebbed with them as Fort Worth, Dallas, and other cities returned to surface water to meet their needs. For a few decades, water levels in the aquifer stabilized, albeit at a lower level than before the aquifer prickled with flowing boreholes. Once pump technology improved and affordable energy became available, production from the aquifer increased yet again, although without the irrational exuberance of the initial discovery wells. Nonetheless, with new production, water levels began to decline once again, this time dropping to more than 300 meters below preproduction levels in the Dallas, Fort Worth, and Waco areas (Mace and others 1994).

Water-level declines in the Trinity Aquifer have been so large that nearly all the artesian pressure in the most heavily produced parts of the aquifer has been exhausted and may have even transitioned the upper

R. E. Mace, *Groundwater Sustainability*, Palgrave Studies in Environmental Sustainability, https://doi.org/10.1007/978-3-031-13516-3_9

artesian formation into water-table conditions in some locales (Mace and others 1994). Since artesian pressures are needed to drive enough water to wells to meet municipal demands, the depletion of artesian pressure affected—and is continuing to affect—the ability of larger wells to continue to produce at previous levels. For example, as water levels have declined in the aquifer, the volume of water supplied to municipal wells in Ellis County south of Dallas has declined by 85 percent over the past 20 years.

Keeping It Safe

At first, human use of groundwater was in harmony with the inflows and outflows of aquifers. But with time, with greater needs and better technology, humans slowly—and then rapidly—began to produce more than what aquifers could provide sustainably. As the consequences of unsustainable production, such as water-level declines, failing springs, seawater intrusion, and land-subsidence, began to be felt, people had to decide if those consequences were acceptable or not, and, if not, how to manage groundwater to prevent them.

When groundwater use was in harmony with inflows and outflows, there was no driving need for groundwater governance. As a result, the default anything-goes approach developed by the Greeks as represented today through the Rule of Capture was arguably appropriate. However, it was not far-sighted, both of human nature and the inexorable march forward of people being able to bend nature to human needs through technology. This was not helped by the hidden, underground nature of groundwater, luring people to assign a magical infiniteness to the resource.

As the technological ability to produce groundwater improved, engineers imported surface-water management techniques to aquifers through the introduction of safe yield. From a sustainability perspective, this was a good start, since safe yield required the ability to produce groundwater perennially. Although initial applications were focused on maximizing production for the sole benefit of human use, safe yield—appropriately implemented—was fundamentally sustainable. The original definition of safe yield as well as subsequent definitions appropriately included a subjective component (such as preventing "dangerous depletion") that required human decisions. As the understanding of aquifers and the impacts of production progressed and the values of society changed, the definition of safe yield also progressed to embrace water quality, economics, and other water users as well as socioeconomics and social welfare, the environment, and future generations.

Unfortunately, many experts and policymakers assume that safe yield is equal to recharge—after all, if you do not spend more than the paycheck that gets deposited into your bank account, you do not go into debt. This Water Budget Myth is easy to adopt since it is easy to think about aquifers (and their water budgets) just like monetary budgets. In many cases, aquifers pumped at their recharge rates will not be pumped sustainably. This may be due to the complexities of getting groundwater into the deeper artesian aquifers, an aquifer leaking into another aquifer, or wanting to maintain springflows and baseflows to surface water bodies.

Although managing groundwater sustainably made sense to engineers, scientists, and academics in the first half of the 1900s, what happened in practice by groundwater producers was often different, leading to unsustainable production. A big reason for this is the long period of time it can take for an aquifer to provide hydrologic feedback to users on unsustainable use. For most of the users of the Ogallala Aquifer, this feedback is multigenerational. Once irrigation investments are made and cultures and economies formed around groundwater irrigation, users are reluctant to reduce their production to sustainable levels. Economic analyses show that, for today's groundwater users, it makes economic sense to produce groundwater in an unsustainable manner, although externalities are often ignored or difficult to quantify and future generations are ignored.

Starting in the 1980s, international discussions turned to sustainable development for the benefit of future generations. Eventually, groundwater became part of these discussions. Although the term sustainable yield had been previously suggested as an alternative term for safe yield, it became more prevalent with the worldwide focus on sustainability. Some claim that sustainable yield and safe yield are two different concepts, but in reality they are the same. Safe yield is saddled with the yoke of history of its technical and policy enlightenment over time, whereas sustainable yield merged into the storyline relatively late. I personally prefer sustainable yield because it gets away from what exactly "safe" means in safe yield (the same concerns as Charles [1946]). Because definitions have changed over time, it is important to make sure everyone agrees on what safe yield, sustainable yield, and groundwater sustainability mean for any particular purpose and what they consider or do not consider.

Through the detailed research I did for this book, I have found the following definitions to be the best:

Groundwater sustainability is the "development and use of ground water in a manner that can be maintained for an indefinite time without causing unacceptable environmental, economic, or social consequences" (Alley and others 1999).

Sustainable yield is how much groundwater can be produced on an annual basis from an aquifer to achieve groundwater sustainability as constrained by unacceptable environmental, economic, or social consequences.

Sustainable production refers to any amount of groundwater production that can be achieved indefinitely.

Safe yield is the same as sustainable yield.

Some experts recommend planning for sustainability over a specified time less than forever, perhaps over a generation. However, sustainable production over a defined period can mask unsustainable production over an infinite period. In my opinion, sustainable means forever. Given that social values may change, the science will (hopefully) improve, and the world is changing (such as climate and land use), planning for forever should be adaptive and allow for periodic reviews and updates to include new information and modify decisions if needed.

At various times, engineers and scientists have taken issue with safe and sustainable yield because of the ambiguity in their definitions, such as "unacceptable environmental, economic, or social consequences" and the question of what is unacceptable. However, any broad discussion of how much groundwater is available for use is, by definition, a policy decision. Policy determines whether an aquifer is managed sustainably or not and what acceptable and unacceptable consequences are. Science informs these policy decisions and quantifies how much groundwater is available for human use once the policy decisions are made. Science informs, people decide.

Governance and management enter the sustainability discussion on how to best manage aquifers and the process to follow to determine if aquifers should be managed sustainably and, if so, what the unacceptable environmental, economic, or social consequences are. Some recent definitions for groundwater sustainability include social justice and equity. While I admire the intent, I believe these processes need to be part of the governance and management discussion and not part of the definition of the sustainability for an aquifer. Adding a social justice component to the definition would result in many aquifers not being considered sustainably managed even though their current governance and management result in sustainable production. From an aquifer's perspective, production is either sustainable or not.

Groundwater sustainability can be attained through a number of means, some easier than others. Some aquifers are protected from unsustainable production by the underlying legal doctrine of its host country or state. Other aquifers are protected by legislation or by court orders. Some aquifers may be produced sustainably without any governance—for the time being—due to low well yields, a land surface unamenable to irrigated agriculture, a low overlying population, or remoteness from thirsty population centers. And yet other aquifers provide hydrologic feedback over the course of a few years, quickly restricting production to sustainable use because of relatively immediate economic impacts that private parties cannot sustain. As Domenico (1972) concluded for managing any aquifer "an inescapable proposition is that use rates must somehow be made to converge on safe yield." Groundwater will be produced sustainably whether we like it or not.

Twice Bitten, Once Shy

Similar to the Leon-Belding Aquifer described in Chap. 9—and for a second time—the Trinity Aquifer has provided hydrologic feedback that it can no longer support current production levels, forcing users to produce less from the aquifer. Also similar to the Leon-Belding Aquifer, decreasing production may result in sustainable production once again. At the end of the day—or, more appropriately, the end of the economic depletion—an aquifer can only give what it can give, no matter how much people want to take.

Some would argue that the groundwater in storage has no value if it is not pumped and that if aquifers will ultimately end up at sustainable production, why not drain them of that water and value before being forced to sustainable development? While this is one possible policy path, the question that has to be asked is: What do we lose along the way? And what is the economic, social, and environmental cost, whether through dollars or intangibles, of that we lose? Users of the Trinity Aquifer have been fortunate in having readily accessible, albeit more expensive, alternative supplies, but users of the Ogallala Aquifer essentially have no monetarily or politically affordable alternatives. As the aquifer dries up, so will the local economy and the food production. The Southern High Plains of Texas will be a very different place in 30 years.

The Texas Commission on Environmental Quality with assessments from the Texas Water Development Board and Texas Parks and Wildlife Department declared the central and northern parts of the Trinity Aquifer as priority groundwater management areas in 2008 and 2009, respectively (TCEQ 2021). A priority groundwater management area means that the state has determined that the area needs groundwater management. In response, locals formed seven groundwater conservation districts to manage the aquifers in North Texas, more than 100 years after problems first arose from producing unsustainably from the aquifer. The future of groundwater sustainability ultimately rests with people and policy.

Questions

Are there opportunities for sustainable groundwater management in your area?

Is there support for managing groundwater sustainably? Who are the advocates?

References

Alley, W.M., Reilly, T.E., and. Franke, O.E., 1999, Sustainability of groundwater resources: U.S. Geological Survey Circular 1186, Denver, Colorado, 79 p.

Charles, J.R., 1946, Discussion on "Utilization of groundwater storage in stream system development": Transactions of the American Society of Civil Engineers, v 111, n 1, p 319–326.

Domenico, P.A., 1972, *Concepts and Models in Groundwater Hydrology*: McGraw-Hill, New York, 405 p.

Mace, R.E., Dutton, A.R., and Nance, H.S., 1994, Water-level declines in the Woodbine, Paluxy, and Trinity aquifers of North-Central Texas: Gulf Coast Association of Geological Societies Transactions, v. 44, p. 413–420.

Salk, J., 1992, Are We Being Good Ancestors? World Affairs: The Journal of International Issues, v. 1, n. 2, p. 16–18.

TCEQ (Texas Commission on Environmental Quality), 2021, Priority groundwater management areas and groundwater conservation districts: Texas Commission on Environmental Quality, SFR-053/20, 28 p.

Glossary

aeolian means windblown.
alluvial means deposited by rivers and streams.
aquifer is geologic media that provides water to wells and springs.
aquifer mining is when water levels are declining because production from the *aquifer* is not sustainable, also referred to as *groundwater mining* or *mining*.
aquifer system is several aquifers closely associated with each other.
aquifer yield is the allowable amount of water that can be produced annually.
artesian aquifer is an aquifer that is under pressure where the water level rises above the top of an aquifer bounded at the top with a confining layer; also a *confined aquifer*.
barrier fault is a fracture with low permeability where groundwater can't flow through easily.
baseflow is flow from an aquifer to stream.
borehole is a narrow shaft sunk into the subsurface.
capillarity refers to the ability for fluids to move in pores due to the forces of adhesion between water and a surface, oftentimes in opposition to gravity.
cementation is the state of sediments such as sand and gravel being hardened through the precipitation of minerals in part of the pore space.
ciénega is a wetland unique to the American Southwest.
cone of depression refers to the three-dimensional shape of water-level declines around a well.

R. E. Mace, *Groundwater Sustainability*, Palgrave Studies in Environmental Sustainability, https://doi.org/10.1007/978-3-031-13516-3

confined aquifer is an aquifer that is under pressure where the water level rises above the top of an aquifer bounded at the top with a confining layer; also an *artesian aquifer*

confined storativity is the volume of water released from storage in a confined aquifer per unit drop in water level (hydraulic head) per unit area of the aquifer. Confined storativity is a function of aquifer and water compressibility.

confining layer is geologic media with a permeability substantially lower than an aquifer such that groundwater flow moves more slowly through it, thus confining the flow of an aquifer; generally also an *aquitard.*

consolidated deposits are geologic media that has been naturally cemented into rock.

cross-formational flow is when groundwater flows from one geologic formation into another one.

depth to water is the distance from the land surface to the water level.

dip refers to the slope of media geologic.

drawdown is the water-level decline in an aquifer due to production.

dynamic equilibrium is when water flowing into and out of aquifers are in a state of balance over time.

economic depletion is where the cost of producing groundwater exceeds the benefit.

equilibrium , with respect to an aquifer, is when flows into, through, and out of an aquifer as well as water levels are not changing over time; opposite of non-*equilibrium.*

evaporation refers to the process of water turning into vapor from liquid water.

evapotranspiration is the sum of evaporation and transpiration from the land surface.

fault is a fracture with displacement.

firm yield is the maximum amount of water that can be pulled out of a reservoir during a repeat of the drought of record.

flowing well is a well where water freely flows from it without the aid of a pump or other human technology to move water.

flowpath is the path water follows in an aquifer.

fracture is a break in geologic media such that a rock is broken into two or more pieces.

grain size is the size of the sand grains and the gravel pieces.

gravitational potential energy is where energy is measured by relative position.

groundwater is water in the saturated zone of geologic media.

groundwater availability is the amount of groundwater available for use.

groundwater governance is "…the decision of which management actions should be taken, when, by who, and for what purpose" (Villholth and Conti 2019).

groundwater management is the implementation of groundwater governance decisions (Villholth and Conti 2019).

groundwater mining is when water levels are declining because production from the *aquifer* is not sustainable, also referred to as *aquifer mining* of mining.

groundwater sustainability is the development and use of groundwater resources in a manner that can be maintained for an indefinite time without causing unacceptable environmental, economic, or social consequences.

heterogeneity refers the variability in geologic media, often leading to variability in permeability.

heterogeneous refers to the state of having heterogeneity.

highly responsive aquifer is an aquifer that responds quickly—on the order of days to months—to changes in recharge and production.

homogeneity refers uniformity in geologic media, often leading to uniformity in permeability.

homogeneous refers to the state of having homogeneity.

hydraulic conductivity describes the ease with which water flows through porous media; it is permeability specific to water.

hydraulic gradient is the change of hydraulic head, as measured through water-level elevations, over a unit distance.

hydraulic head is a measure of the potential energy of water, generally measured through water levels; also referred to as *water-level elevation*.

infiltration is the movement of water from the land surface into the ground.

intrinsic permeability is the measure of the ability for porous media to transmit fluids; also *permeability*.

isotropy is when permeability is the same no matter which direction it is measured in.

joint is a fracture with movement along the fracture plane.

karst refers to partially dissolved limestones.

limestone is a rock created from the precipitation of calcite and the accumulation of calcareous debris from algae and animals.

maximum sustainable yield is the annual amount of water that can be produced from an aquifer in perpetuity, given current climate conditions, without consideration of undesirable outcomes.

mining is when water levels are declining because production from the aquifer is not sustainable, also referred to as *aquifer mining* or *groundwater mining*.

monitoring well is a borehole used to monitor an aquifer.

non-equilibrium , with respect to an aquifer, is when flows into, through, and out of aquifer as well or water levels are changing over time; opposite of *equilibrium*.

noria , see *waterwheel*.

percolation is the movement of water through the soil after infiltration.

permeability describes the ease with which fluids flow through porous media; it is independent of fluid type. It is often used as a general term to describe the ability of geologic media to transmit fluids regardless of the fluid.

phreatophyte is a plant that has roots that reach below the water table and into an aquifer.

policy is "a set of ideas or a plan of what to do in particular situations that has been agreed to officially by a group of people, a business organization, a government, or a political party" (Cambridge Dictionary 2021).

pores are the spaces between the solid particles that make up soil, sediment, and rocks.

porosity is the fraction that pores make-up of a volume of porous media.

porous media are geologic materials such as sand, clay, sandstone, limestone, and shale that have pores.

potentiometric surface is the contoured elevations of water levels or hydraulic heads in geologic media.

qanat is a gently sloping tunnel that intersects the water table at a higher elevation—generally in alluvium along a mountain front—and carries the water by gravity to a lower elevation.

recharge is water that percolates to and crosses the water table.

renewability is the natural average amount of water that enters the watershed each year.

safe yield (for surface water) includes a drought worse than a drought of record, usually defined as providing the yield for a certain amount of time longer.

safe yield (for groundwater) is the volume of water that can be perennially produced from an aquifer without producing undesirable results.

saline is possessing salts and/or solids.

sandstones are *consolidated deposits* of sand.

saturated thickness is the vertical distance between the bottom of an aquifer and the top of the aquifer.

saturated zone is geologic media that is fully saturated with water.

semi-confined aquifer is an aquifer that is confined but the confining layer can transmit a substantial amount of water.

shaduf is an early machine used to access groundwater in wells and is based on a simple lever mechanism with a counterweight to the water.

slowly responsive aquifer is an aquifer that does not respond quickly to changes in recharge and production, on the order of decades or more.

spring is a natural feature where groundwater appreciably discharges to the land surface.

static water level is the water level in the aquifer without being affected by local groundwater production.

steady state when an aquifer is in equilibrium.

strong sustainability where there is no substitute for the water.

sustainable is able to be maintained in perpetuity without interruption or weakening.

sustainable development is development that meets the needs of the present without compromising the ability of future generations to meet their own needs.

sustainable production is the amount of production from an aquifer that can maintained indefinitely.

tragedy of the commons is where the system compels each individual to increase their use of a shared but limited resource.

transition zone is where an aquifer is unconfined when water levels are low and confined when water levels are high, generally tied to seasons or changes in recharge rates.

transpiration is the use of water by plants.

tympanon is a water wheel that uses a chain of buckets instead of buckets attached to the wheel.

unconfined aquifer is an aquifer where the top of the aquifer is defined by a water table; also a *water-table aquifer*.

unconfined storativity is the volume of water released from storage in an unconfined aquifer per unit drop in water level (hydraulic head) per unit area of the aquifer. Unconfined storativity is unitless and is the same as the drainable porosity of an aquifer.

unconsolidated deposits are geologic media that has not been naturally cemented into rock.

unsaturated zone is where the geologic media where there is air in the subsurface; also the *vadose zone*.

vadose zone is where the geologic media where there is air in the subsurface; also the *unsaturated zone*.

virga is rainfall that evaporates before hitting the ground.

water level is a measure of the potential energy in a part of an aquifer; the level to which water would rise in a well.

water-level elevation is the elevation of the water level above a benchmark, usually mean sea level.

water table is the top of an unconfined aquifer.

water-table aquifer is an aquifer where the top of the aquifer is defined by a water table; also an *unconfined aquifer*.

waterwheel is a horizontal axis wheel powered by water flow that also captures water in a series of rotating buckets attached to the wheel.

weak sustainability assumes that the generational hand-off will include new and more efficient technology to use water.

well is a borehole used to extract oil, gas, or water or for other purposes, such as monitoring.

yield is a term that quantifies the capacity of a water resource to serve as a long-term water supply.

References

Adhikari, S., and Ivins, E.R., 2016, Climate-driven polar motion – 2003–2015: Science Advances, 2: e1501693, 10 p.

Aeschbach-Hertig, W., & Gleeson, T., 2012, Regional strategies for the accelerating global problem of groundwater depletion. Nature Geoscience, v 5, n 12, p. 853–861.

Agricola, G., 1556, De re metallica: 1st edition, Basil, Hieronymus Froben & Nicolaus Episcopius, 594 p.

Ágústsdóttir, A.M., 2015, Ecosystem approach for natural hazard mitigation of volcanic tephra in Iceland – Building resilience and sustainability: Natural Hazards, v. 78, p. 1669–1691.

Aiken, J.D., and Supalla, R.J., 1979, Ground water mining and western water rights law – The Nebraska experience: South Dakota Law Review, v 24, p. 607–648.

Allan, P., 1906, The drought antidote for the North-West: Or, the utilisation of the Artesian Resources of New South Wales: Sydney, Australia, Government Printer.

Allen, M., 1915, Discussion of "The determination of safe yield of underground reservoirs of the closed basin type:" Transactions of the American Society of Civil Engineers, v 78, n 1, p 239–242.

Alley, W.M., and Leake, S.A., 2004, The journey from safe yield to sustainability: Ground Water, v 49, n 1, p 12–16.

R. E. Mace, *Groundwater Sustainability*, Palgrave Studies in Environmental Sustainability, https://doi.org/10.1007/978-3-031-13516-3

Alley, W.M., Reilly, T.E., and Franke, O.E., 1999, Sustainability of groundwater resources: U.S. Geological Survey Circular 1186, Denver, Colorado, 79 p. https://pubs.usgs.gov/circ/circ1186/.

American Journal of Science, 1853, v. XV.

Anderson, D.A., 2019, Environmental Economics and Natural Resource Management, 5th Edition: Routledge, New York, 446 p.

Anderson, M.P., and Berkebile, C.A., 1977, Hydrogeology of the South Fork of Long Island, New York – Discussion: Geological Society of America Bulletin, v. 88, p. 895.

Angula, H.K., 2001, Foreword *in* Christelis, G., and Struckmeier, W., Groundwater in Namibia – An explanation to the hydrogeological map: Ministry of Agriculture, Water, and Rural Development, Namibia, p 2.

Arago, F., 1834, Sur les puits forés, connus sous le nom de puits artésiens, des fontaines artésiennes, ou de fontaines jaillissants: Bureau des Longitudes, Annuaire pour l'an 1835, Paris, p 181–258.

ARMCANZ (Agriculture and Resource Management Council of Australia and New Zealand), 1996, Allocation and use of groundwater – A national framework for improved groundwater management in Australia: Occasional Paper No. 2, Commonwealth of Australia.

Arnadóttir, N., 2014, E. Coli found in drinking water: Grapevine, published September 9, 2014.

ASCE (American Society of Civil Engineering), 1961, Ground water basin management: Manual of Engineering Practice No. 40, 160 p. https://babel.hathitrust.org/cgi/pt?id=coo.31924000122311&view=1up&seq=8.

ASCE Task Committee on Sustainability Criteria. 1998. Sustainability Criteria for Water Resource Systems. American Society for Civil Engineers, Reston, Virginia, USA. 253 pages.

AWIC (Artesian Water Investigation Committee), 1945, Artesian water supplies, First Interim Report: Queensland Artesian Water Investigation Committee, Queensland, 40 p.

AWIC (Artesian Water Investigation Committee), 1954, Artesian water supplies in Queensland: Queensland Artesian Water Investigation Committee, Department of the Co-ordinator-General of Public Works, Queensland, 79 p.

Axelssona, G., Gunnlaugsson, E., Jónasson, T., and Ólafsson, M., 2010, Low-temperature geothermal utilization in Iceland – Decades of experience: Geothermics, v. 39, p. 329–338.

Bagheri-Gavkosh, M., Hosseini, S.M., Ataie-Ashtiani, B., Sohani, Y., Ebrahimian, H., Morovat, F., and Ashrafi, S., 2021, Land subsidence – A global challenge: Science of the Total Environment, https://doi.org/10.1016/j.scitotenv.2021.146193, 17 p.

Bagley, E.S., 1961, Water rights, law and public policy relating to ground water "mining" in the Southwestern States: The Journal of Law & Economics, v. 4, p. 144–174.

Bagley, E.S., 1972, The Ground-Water Depletion Allowance under the Federal Income Tax: Natural Resources Journal, v 12, n 3, p 445–466.

Baker, D.M., 1946, Discussion on "Utilization of groundwater storage in stream system development": Transactions of the American Society of Civil Engineers, v 111, n 1, p 343–346.

Baker, T.L., 1985, A Field Guide to American Windmills: University of Oklahoma Press, 516 p.

Banks, H.O., 1953, Utilization of underground storage reservoirs: Transactions of the American Society of Civil Engineers, v 118, p 220–234.

Bartlett, J., 1862, Basset v. Salisbury Manufacturing Company – Decision: New Hampshire Supreme Court, 43 New Hampshire 328.

Barnett, S., Simmons, C.T., and Nelson, R., 2021, Groundwater resources in Australia – their occurrence, management, and future challenges *in* Mukherjee, A., Scanlon, B.R., Aureli, A., Langan, S., Guo, H., and McKenzie, A.A., editors, Global Groundwater – Source, Scarcity, Sustainability, Security, and Solutions: Chapter 3, Elsevier, Amsterdam, Netherlands, p. 35–46.

Barraqué, B., and de Marsily, G., 2010, Country report for France *in* Groundwater in the Southern Member States of the European Union – An assessment of current knowledge and future prospects: European Academies Science Advisory Council, EASAC policy report 12, SBN: 978-3-8047-2827-1, German Academy of Sciences, 44 p.

Barrick, K.A., 2010, Environmental review of geyser basins – Resources, scarcity, threats, and benefits: Environment Reviews, v. 18, p. 209–238.

Bear, J., Cheng, A.H.-D., Sorek, S., Ouazar, D., and Herrera, I., (editors), 1999, Seawater intrusion in coastal aquifers – Concepts, Methods, and Practices: Kluwer Academic Publishers, 627 p.

Bear, J., and Levin, O., 1967, The optimal yield of an aquifer: International Association of Scientific Hydrology Bulletin, Publication 72, p. 401–412.

Beaumont, P., 1985, Irrigated agriculture and ground-water mining on the High Plains of Texas, USA: Environmental Conservation, v 12, n 2, p 119–130.

Bell, T.J., 1881, History of the water supply of the world: Cincinnati, Ohio, Peter G. Thomson, Publisher, 134 p.

Blake, T., and Cook, M., 2006, Great Artesian Basin historical overview: Queensland Department of Natural Resources Mines and Water, Brisbane, Australia

Blumberg, H., and Collins, G., 2016, Implementing three-dimensional groundwater management in a Texas groundwater conservation district: Texas Water Journal, v. 7, n. 1, p. 69–81.

Blumm, M.C., and Ritchie, L., 2005, The pioneer spirit and the public trust – The American Rule of Capture and state ownership of wildlife: Environmental Law, v. 35, n. 4, p. 101–147. https://ssrn.com/abstract=845044.

Bobeck, P., 2006, Henry Darcy in his own words: Hydrogeology Journal, v 14, p 998–1004.

Bodvarsson, G., and Zais, E., 1978, A field example of free surface testing: in Kruger, P., and Ramey, Jr., H.J., editors, Proceedings Forth Workshop Geothermal Reservoir Engineering, SGP-TR-30, CONF-781222-20, Stanford, California, p. 153–159.

Bollig, M., 2013, Social-ecological change and institutional development in a pastoral community in North-Western Namibia *in* Boling, M., Wotzka, H.P., and Schnegg, M., editors, Pastoralism in Africa – Past, present and futures: London, Berghahn, p. 316–340.

Brake, L, 2020, Development, management and rehabilitation of water bores in the Great Artesian Basin, 1878–2020: Proceedings of The Royal Society of Queensland, v 126, p 153–175.

Bredehoeft, J.D., 1997, Safe yield and the water budget myth: Ground Water, v. 35, n. 6, p. 929.

Bredehoeft, J.D. 2002. The water budget myth revisited: Why hydrogeologists model. Ground Water, v. 40, n. 4, p. 340–345.

Bredehoeft, J.D., Papadopulos, S.S., and Cooper, H.H., 1982, Groundwater – The water budget myth: *in* Scientific Basis of Water-Resource Management, Studies in Geophysics, Washington, DC: National Academy Press, p. 51–57.

Brinkmann, R., 2020, *Environmental Sustainability in a Time of Change*: Palgrave Studies in Environmental Sustainability, Palgrave Macmillan, 325 p.

Brown, L.R., 1982, Building a sustainable society: Society, v19, p 75–85.

Brune, G., 1975, Major and historical springs of Texas: Texas Water Development Board Report 189, 94 p. + plates.

Brune, G., 1981, Springs of Texas, Volume 1: Branch-Smith, Inc., Fort Worth Texas, 566 p.

BSEACD (Barton Springs Edwards Aquifer Conservation District), 2022, History: accessed January 4, 2022; https://bseacd.org/about-us/history/.

Buchanan, J.M., 1959, Positive economics, welfare economics, and political economy: Journal of Law and Economics, v. 2, n. 2, p. 124–138.

Bullard, L., 2004, Healing waters – Missouri's historic mineral springs and spas: University of Missouri Press, 243 p.

Bumgarner, J.R., Stanton, G.P., Teeple, A.P., Thomas, J.V., Houston, N.A., Payne, J.D., and Musgrove, M., 2012, A Conceptual Model of the Hydrogeologic Framework, Geochemistry, and Groundwater-Flow System of the Edwards-Trinity and Related Aquifers in the Pecos County Region, Texas: Scientific Investigations Report 2012–5124 (revised July 10, 2012), U.S. Geological Survey, 74 p.

Burt, O.R., 1964, Optimal resource use over time with an application to ground water, Management Science, v. 11, p. 80–93.

Burt, O.R., 1967, Temporal allocation of groundwater, Water Resources Research, v. 3, no. 1, p. 45–56.

Byrne, D., Eno, B., Frantz, C., Harrison, J., and Weymouth, T., 1980, Once in a Lifetime: *in* Remain in Light, Track 1, Side 2, Sire Records.

Calderone, L., 2018, Using windmills to deliver water: AgriTech Tomorrow, posted March 13, 2018, https://www.agritechtomorrow.com/article/2018/03/using-windmills-to-deliver-water/10595/.

Cambridge Dictionary, 2021a, Policy: https://dictionary.cambridge.org/dictionary/english/policy accessed October 1, 2021.

Canadell, J., Jackson. R.B., Ehleringer, J.B., Mooney, H.A., Sala, O.E., and Schulze, E.D., 1996, Maximum rooting depth of vegetation types at the global scale: Oecologia, v 108, n 4, p 583–595.

Carlyle, T., 1849, Occasional discourse on the Negro question: Fraser's Magazine for Town and Country, v 40, p 670–679.

CDWR (California Department of Water Resources), 2020a, California's groundwater and the Sustainable Groundwater Management Act: January 2020, accessed on December 27, 2021: https://water.ca.gov/-/media/DWR-Website/Web-Pages/Programs/Groundwater-Management/Sustainable-Groundwater-Management/Files/CA-Groundwater-and-SGMA-Fact-Sheet.pdf.

CDWR (California Department of Water Resources), 2020b, Sustainable Groundwater Management Act 2019 basin prioritization – Process and results: May 2020, variously paginated (99 total pages).

CDWR (California Department of Water Resources), 2021a, Groundwater: accessed on December 5, 2021, https://water.ca.gov/water-basics/groundwater.

CDWR (California Department of Water Resources), 2021b, Sustainable Groundwater Management Act – Plan evaluation pathways: May 2021, accessed on December 27, 2021, https://water.ca.gov/-/media/DWR-Website/Web-Pages/Programs/Groundwater-Management/Sustainable-Groundwater-Management/Groundwater-Sustainability-Plans/Files/GSP/SGMA-Evaluation-Pathways-Factsheet.pdf.

CG-ASCE (Committee on Groundwater for the American Society of Civil Engineers), 1961, Ground water basin management: Manual of Engineering Practice No. 40, 160 p.

Charalabidis, Y., Zuiderwijk, A., Alexopoulos, C., Janssen, M., Lampoltshammer, T., and Ferro, E., 2018, The World of Open Data – Concepts, Methods, Tools and Experiences: Springer Nature, 229 p.

Charles, J.R., 1946, Discussion on "Utilization of groundwater storage in stream system development": Transactions of the American Society of Civil Engineers, v 111, n 1, p 319–326.

Chondros, T.G., 2010, Archimedes influence in science and engineering in Paipetis, S.A., and Ceccarelli, M., editors, The Genius of Archimedes – 23 Centuries of Influence on Mathematics, Science and Engineering, History of Mechanism and Machine Science, Springer Science+Business Media, p. 411–425, https://doi.org/10.1007/978-90-481-9091-1_31.

Christelis, G., Heyns, P., and Struckmeier, W., 2001, Essentials of groundwater *in* Christelis, G., and Struckmeier, W., Groundwater in Namibia – An explanation to the hydrogeological map: Ministry of Agriculture, Water, and Rural Development, Namibia, p 30–39.

Christelis, G., and Struckmeier, W., 2001, Groundwater in Namibia – An explanation to the hydrogeological map: Ministry of Agriculture, Water, and Rural Development, Namibia, 128 p.

Clark, G., 1944, Antiquity: v 18, n 69, p 1–15.

CNRA, CEPA, and CDFA (California Natural Resources Agency, California Environmental Protection Agency, and California Department of Food & Agriculture), 2013, California agencies release draft action plan for water, ask for input and dialogue – California Water Action Plan provides roadmap for state efforts: press release, October 31. https://www.waterboards.ca.gov/press_room/press_releases/2013.shtml.

CNRA, CEPA, and CDFA (California Natural Resources Agency, California Environmental Protection Agency, and California Department of Food & Agriculture), 2014, California water action plan: joint publication, 19 p., https://resources.ca.gov/CNRALegacyFiles/docs/california_water_action_plan/2014_California_Water_Action_Plan.pdf.

COA (City of Austin), 2022, Barton Springs Pool: accessed on January 4, 2022; https://www.austintexas.gov/department/barton-springs-pool#:~:text=Depths%20of%20the%20pool%20range,Edwards%20Aquifer%20which%20feeds%20it.

CoAG (Council of Australian Governments), 1994. Communique of February Meeting. Attachment A. National Competition Council, http://ncp.ncc.gov.au/docs/PIAg-001.pdf.

Collins, G., 2021, Overruling the Rule of Capture – What can Texas learn from 10 other states' groundwater law updates: Center for Energy Studies at Rice University's Baker Institute for Public Policy and The Meadows Center for Water and the Environment at Texas State University, 56 p., https://doi.org/10.25613/knsb-6g39.

Combes, M., 1838, Surles Roues de Reaction: Académie des Sciences, v 7.

Combes, M., 1840, Recherches expérimentales sur les roues à réaction: Académie des Sciences, v 10, p 362.

Conkling, H., 1946, Utilization of groundwater storage in stream system development: Transactions of the American Society of Civil Engineers, v 111, n 1, p 33–62.

Conner, C.D., 2005, A people's history of science: Nation Books, 554 p.

Conroy, M.J., and Peterson, J.T., 2013, Decision Making in Natural Resource Management – A Structured, Adaptive Approach: Wiley-Blackwell, New York, 480 p.

Constitute, 2021, Namibia's Constitution of 1990 with Amendments through 2010: constitureproject.com, 68 p., accessed April 20, 2022.

Contoux, C., Violette, S., Vivona, R., Goblet, P., and Patriarche, D., 2013, How basin model results enable the study of multi-layer aquifer response to pumping: the Paris Basin, France: Hydrogeology Journal, v. 21, p. 545–557.

Cox, R., and Barron, A., editors, 1998, Great Artesian Basin resource study: Great Artesian Basin Consultative Council, 235 p.

Cox, W.G., 1895, Artesian wells as a means of supply: Sapsford & Company, Brisbane, Australia, 146 p.

Curtis, I., Hook, P., Morris, S., and Sumner, B., 1979, From Safety to Where…? *in* EARCOM 2 – Contradition, Fast Product, Edinburgh, England.

Dafny, E., 2016, The Great Artesian Basin – Is it that great? Hydrogeology Journal, https://doi.org/10.1007/s10040-016-1444-5.

Dallas Morning News, October 15, 1890, The Dallas Geyser.

Dallas Morning News, April 26, 1891, Dallas boring wells.

Dallas Morning News, January 4, 1892, Artesian investigation – Condensation of Prof. R.T. Hill's forthcoming report.

Dallas Morning News, August 24, 1893, Artesianwater – The great lake of it on which Texas rests.

Darcy, H., 1856, Les Fontaines Publiques de la Ville de Dijon.

DAWE (Department of Agriculture, Water and the Environment), 2020, Great Artesian Basin Strategic Management Plan: Department of Agriculture, Water and the Environment, Australian Government, 53 p.

Degousée, J., 1847, Guide du sondeur, ou Traité théorique et pratique des sondages: Langlois et Leclercq, Paris, Volume 1, 504 p.

De Feo, G., Angelakis, A.N., Antoniou, G.P., El-Gohary, F., Haut, B., Passchier, C.W., and Zheng, X.Y., 2013, Historical and technical notes on aqueducts from prehistoric to medieval times: Water, v 5, p 1996–2025; https://doi.org/10.3390/w5041996.

de Saint-Exupery, A., 1943, The Little Prince: Reynal and Hitchcock, New York.

Deal, K.H., 2016, Wildlife and Natural Resource Management, 4th Edition: Delmar Cengage Learning, Boston, 340 p.

Dellapenna, J.W., 2013, A primer on groundwater law: Idaho Law Review, v. 49, p. 265–323.

Deutsches Museum, 2021, Roman double-piston pump: accessed April 11, 2021, https://www.deutsches-museum.de/en/collections/machines/power-engines/water-power/piston-pumps/double-piston-pump/.

Devlin, J.F., Sophocleous, M., 2004, The persistence of the water budget myth and its relationship to sustainability. Hydrogeology Journal, v. 13, 549–554.

DEWHA (Department of the Environment, Water, Heritage and the Arts), 2009, Water Matters: Australian Government, Issue 4, May, 1 p.

Dickinson, R., Pavley, F., and Rendon, A., 2014, Groundwater management: Assembly Bill No. 1739, https://leginfo.legislature.ca.gov/faces/billTextClient.xhtml?bill_id=201320140AB1739.

Dillon, P., Escalante, E.F., Megdal, S.B., and Massmann, G., editors, 2021, Managed aquifer recharge for water resilience: Water, Special Issue, 408 p. ISBN 978-3-03943-043-7

Dochartaigh, B.É.Ó., MacDonald, A.M., Black, A.R., Everest, J., Wilson, P., Darling, W.G., Jones, L., and Raines, M., 2019, Groundwater–glacier meltwater interaction in proglacial aquifers: Hydrology and Earth System Sciences, v. 23, no. 11, p. 4527-4539 10.5194/hess-23-4527-2019

DoE (Department of the Environment), 2014, Great Artesian Basin sustainability initiative (GABSI) value for money review: Department of the Environment, Australia, 52 p.

Döll, P., Trautmann, T., Göllner, M, and Schmied, H.M., 2019, A global-scale analysis of water storage dynamics of inland wetlands – Quantifying the impacts of human water use and man-made reservoirs as well as the unavoidable and avoidable impacts of climate change: Ecohydrology, 18 p., https://doi.org/10.1002/eco.2175.

Domenico, P.A., 1972, *Concepts and Models in Groundwater Hydrology*: McGraw-Hill, New York, 405 p.

Domenico, P.A., Anderson, D., and Case, C., 1968, Optimal ground-water mining: Water Resources Research, v 4, n 2, p 247–255.

Domenico, P.A., and Schwartz, F.W., 1990, Physical and Chemical Hydrogeology: John Wiley & Sons, 824 p.

Downing, R.A., 1986, Development of groundwater *in* Brandon, T.W., editor, *Groundwater Occurrence, Development and Protection*: Institution of Water Engineers and Scientists, London, p 485–542.

Downing, R.A., 1993, Groundwater resources, their development and management in the UK – An historical perspective: Quarterly Journal of Engineering Geology, v 26, p 335–358.

Drummond, G.B., 1946, Discussion on "Utilization of groundwater storage in stream system development": Transactions of the American Society of Civil Engineers, v 111, n 1, p 337–338.

Drummond, D.O., Sherman, L.R., and McCarthy, E.R., Jr., 2004, The Rule of Capture in Texas – Still so misunderstood after all these years: Texas Tech Law Review, v 37, n 1, 97 p.

Dupaigne, T., and, Violette, S., 2020, Impact de l'activité anthropique et des règles de gestion des eaux souterraines sur la ressource – exemple de l'Albien en Île-de-France: La Houille Blanche, 106:1, 16–21, https://doi.org/10.1051/lhb/2020003.

EUWD (Edwards Underground Water District), 1989, Edwards Underground Water District created to conserve, protect, and preserve the Edwards Aquifer: Edwards Underground Water District, 22 p.

Evans, W.R., and Cook, P.G., 2002, What is a sustainable yield for Australia's groundwater systems? International Association of Hydrogeologists Darwin Conference: Balancing the Groundwater Budget, Darwin, Australia, May 12–17.

Fairfax R.J., and Fensham, R.J., 2002, In the footsteps of J. Alfred Griffiths – A cataclysmic history of Great Artesian Basin springs in Queensland: Australian Geographical Studies, v 40, n 2, p 210–230.

Falk, T., Bock, B., and Kirk, M., 2009, Polycentrism and poverty – Experiences of rural water supply reform in Namibia: Water Alternatives, v 2, p 115–137.

Famiglietti, J.S., 2014, The global groundwater crisis: Nature Climate Change, v 4, p 945–948.

Fattahi, M., 2015, OSL dating of the Miam Qanat (KARIZ) system in NE Iran: Journal of Archaeological Science, v 59, p. 54–63.

Fensham, R., and Fairfax, R., 2005, Great Artesian Basin water resource plan ecological assessment of GAB springs in Queensland: Department of Natural Resources and Mines, Brisbane, Australia, 27 p.

Fetter, Jr., C.W., 1972, The concept of safe groundwater yield in coastal aquifers: Water Resources Bulletin, v. 8, n. 6, p. 1173–1176.

Fetter, C.W., 1980, Applied Hydrogeology (1st Edition): Prentice-Hall, 544 p.

Fetter, C.W., 2001, Applied Hydrogeology (4th Edition): Prentice-Hall, 598 p.

Fisher, N.H., and Hahn, G.W., 1963, Review of available groundwater data on the Great Artesian Basin: Bureau of Mineral Resources, Geology and Geophysics. Records 1963/128, variously paginated (20 total pages).

Foley, J. A., Ramankutty, N., Brauman, K.A., Cassidy, E.S., Gerber, J.S., Johnston, M., Mueller, N.D., O'Connell, C., Ray, D.K., West, P.C., Balzer, C., Bennett, E.M., Carpenter, S.R., Hill, J., Monfreda, C., Polasky, S., Rockstrom, J., Sheeha, J., Siebert, S., Tilman, D., and Zaks, D.P.M, 2011, Solutions for a cultivated planet: Nature, v 478, 337–342.

Forbes, H., 1946, Discussion on "Utilization of groundwater storage in stream system development": Transactions of the American Society of Civil Engineers, v 111, n 1, p 346–338.

Fort Worth Gazette, 1891, Artesian Water-Fort Worth Lays Claim to Being the Geyser City-Three Powerful Artesian Veins Underneath the City-Five Hundred Gallons Per Minute Sixty Feet Above the City: v 15, n 222, edition 1, Monday, May 25, 1891.

Fox, P., editor, 2007, Management of aquifer recharge for sustainability: Acacia Publishing, 641 p.

Foster, S., Garduño, H., Tuinhof, A., and Tovey, C., 2009, Groundwater Governance: Strategic Overview Series No. 1. Washington, D.C.: World Bank.

Foster, S., and Tyson, G., 2019, Climate-change adaptation & groundwater: Strategic Overview Series, International Association of Hydrogeologists, 6 p.

Foxx, T.S., Tierney, G.D., and Williams, J.M., 1984, Rooting depths of plants relative to biological and environmental factors: Los Alamos National Laboratory report, 26 p.

Franzson, H., Tómasson, J., Sverrisdóttir, G., and Sigurosson, F., 1997, Geothermal energy in Iceland: *in* Mineral and Thermal Groundwater Resources. Springer, Dordrecht, Chapter 9. https://doi.org/10.1007/978-94-011-5846-6_9.

Freeze, R.A., 1971, Three-dimensional, transient, saturated-unsaturated flow in a groundwater basin: Water Resources Research, v. 7, n. 2, p. 347–366.

Freeze, R.A., and Cherry, J.A., 1979, Groundwater: Prentice-Hall, Inc., 604 p.

Frontier Economics, 2016, Economic output of groundwater dependent sectors in the Great Artesian Basin: A report commissioned by the Australian Government and Great Artesian Basin Jurisdictions based on advice from the Great Artesian Basin Coordinating Committee, Melbourne, Australia, 60 p.

GABCC (Great Artesian Basin Consultative Council), 2000, Great Artesian Basin Strategic Management Plan: Great Artesian Basin Consultative Council for the Australian Government Department of Agriculture, Fisheries and Forestry, Canberra, 60 p.

Gardner, M., 2017, Happy BD to BSEACD! Conservation district celebrates 30 years: Hays Free Press, p. 1C–2C.

Garfinkel, Y., Vered, A., and Bar-Yosef, O., 2006, The domestication of water – The Neolithic well at Sha'ar Hagolan, Jordan Valley, Israel: Antiquity, v 80, p 686–96.

Gaye, C.B., and Tindimugaya, C., 2019, Review – Challenges and opportunities for sustainable groundwater management in Africa: Hydrogeology Journal, v 27, p 1099–1110. https://doi.org/10.1007/s10040-018-1892-1.

Geirsson, H., Árnadóttir, T., Völksen, C., Jiang, W., Sturkell, E., Villemin, T., Einarsson, P., Sigmundsson, F., and Stefánsson, R., 2006, Current plate movements across the Mid-Atlantic Ridge determined from 5 years of continuous GPS measurements in Iceland: Journal of Geophysical Research – Solid Earth, https://doi.org/10.1029/2005JB003717.

Giller, G., 2017, A fragile balance – The endangered Barton Springs salamanders of Texas: Reptiles, accessed January 4, 2022; https://reptilesmagazine.com/a-fragile-balance-the-endangered-barton-springs-salamanders-of-texas/.

Gleeson, T., Alley, W.M., Allen, D.M., Sophocleous, M.A., Zhou, Y., Taniguchi, M., and VanderSteen, J., 2012, Towards sustainable groundwater use: Setting long-term goals, backcasting, and managing adaptively: Groundwater, v. 50, n. 1, p. 19–26.

Gleeson, T., Cuthbert, M., Ferguson, G., and Perrone, D. 2020, Global groundwater sustainability, resources, and systems in the Anthropocene: Annual Review of Earth and Planetary Science, v. 48, p. 431–463.

Gleeson, T., Villholth, K., Taylor, R., Perrone, D., Famiglietti, J., Hyndman, D., van Loon, A., and Wada, Y., (leadership team), 2019, Global groundwater sustainability – A call to action: Global Groundwater Statement, groundwaterstatement.org, 2 p.

Global Environment Facility, 2015, Global Diagnostic on Groundwater Governance: Special Edition for World Water Forum 7, 186 p. https://groundwaterportal.net/resource/global-diagnostic.

Goes, B.J.M., Parajuli, U.N., Haq, M., and Wardlaw, R.B., 2017, Karez (qanat) irrigation in the Helmand River Basin, Afghanistan – A vanishing indigenous legacy: Hydrogeology Journal, v 25, p 269–286.

Green, D.E., 1973, Land of the Underground Rain – Irrigation on the Texas High Plains, 1910–1970: University of Texas Press, 295 p.

Griffin, R.C., 2006, Water resource economics – The analysis of scarcity, policies, and projects: MIT Press, Cambridge, Massachusetts, 496 p.

Griffin, J.M., 2017, Interjecting economics into the groundwater policy dialogue: Texas Water Journal, v. 8, n. 1, p. 97–112.

Griggs, B.W., 2021, Reaching consensus about conservation: High Plains lessons for California's Sustainable Groundwater Management Act: University of the Pacific Law Review, v. 52, n. 3, p. 495–547.

Gulley, R.L., 2015, Heads above water – The inside story of the Edwards Aquifer Recovery Implementation Program: Texas A&M University Press, 234 p.

Gunnarsdottir, M.J., Gardarsson, S.M., and Andradottir, H.O, 2013, Microbial contamination in groundwater supply in a cold climate and coarse soil – Case study of norovirus outbreak at Lake Mývatn, Iceland: Hydrology Research, 44.6, p. 1114–1128.

Gunnarsdottira, M.J., Gardarsson, S.M., and Bartram, J., 2014, Developing a national framework for safe drinking water – Case study from Iceland: International Journal of Hygiene and Environmental Health, IJHEH-12804. https://doi.org/10.1016/j.ijheh.2014.10.003.

Gunnarsdottir, M.J., Gardarsson, S.M., Jonsson, G.S., Armannsson, H., and Bartram, J., 2015, Natural background levels for chemicals in Icelandic aquifers: Hydrology Research, 46.4, p. 647–660.

Gunnlaugsson, E., and Ívarsson, G., 2010, Direct use of geothermal water for district heating in Reykjavík and other towns and communities in SW-Iceland: Proceedings of the World Geothermal Congress 2010, Bali, Indonesia, 6 p.

Habermehl, R.A., 1980, The Great Artesian Basin, Australia: BMR Journal of Australian Geology &: Geophysics, v 5, p 9–38.

Habermehl, R.A., 2018, Groundwater governance in the Great Artesian Basin, Australia *in* Villholth, K.G., López-Gunn, E., Conti, K.I., Garrido, A., and van der Gun, J., Advances in Groundwater Governance: CRC Press, Taylor & Francis Group, p 411–441.

Hallenburg, J.K., 1997, Introduction to geophysical formation evaluation: CRC Press, 181 p.

Hamlin, S.N., and Anthony, S.S., 1987, Ground-water resources of the Laura Area, Majuro Atoll, Marshall Islands: U.S. Geological Survey, Water Resources Investigations Report 87-4047, 69 p.

Hardin, G., 1968, The tragedy of the commons: Science, v 162, n 3859, p 1243–1248.

Harris, C., 2002, Culture and geography: South Australia's mound springs as trade and communication routes: Historic Environment, v 16, n 2, p 8–11.

Harris, L.E., 1953, Some factors in the early development of the centrifugal pump 1689 to 1851: Transactions of the Newcomen Society, v 28, n 1, p 187–202, https://doi.org/10.1179/tns.1951.014.

Hartman, L.M., 1965, Economics and groundwater development: Ground Water, v 2, n 2, p 4–8.

Heyns, P., Seely, M., and Montgomery, S., 1998, Namibia's water – A decision maker's guide: Ministry of Agriculture, Water and Rural Development, Namibia, 173 p.

Heyns, P., and Struckmeier, W., 2001, Introduction – Water in Namibia and rationale for a hydrogeological map project *in* Christelis, G., and Struckmeier, W., Groundwater in Namibia – An explanation to the hydrogeological map: Ministry of Agriculture, Water, and Rural Development, Namibia, p 9–12.

Hills, R.L., 1994, Power from Wind – A History of Windmill Technology: Cambridge University Press, 324 p.

Hill, R.T., 1892, On the occurrence of artesian and other underground water in Texas, New Mexico, and Indian Territory together with the geology and geography of those regions: Final Reports of the Artesian and Underflow Investigations of the Department of Agriculture, Washington, D.C., 166 p.

Hill, R.T., 1901, Geography and geology of the Black and Grand Prairies, Texas, with detailed descriptions of the Cretaceous formations and special reference to artesian waters: Twenty-First Annual Report of the United States Geological Survey to Secretary of the Interior, 1899–1900, Part VII-Texas, Government Printing Office, Washington, 666 p. + plates.

Hill, R.T., and Vaughan, T.W., 1898, Geology of the Edwards Plateau and Rio Grande Plain adjacent to Austin and San Antonio, Texas, with reference to the occurrence of underground waters: in Walcott, C.D., (director), Eighteenth Annual Report of the United State Geological Survey to the Secretary of the Interior, 1896–97, Part II – Papers Chiefly of a Theoretic Nature, Government Printing Office, Washington, D.C., p. 193–322.

Hinz, M.O., 2016, Customary law ascertained – The customary law of the Nama, Ovaherero, Ovambanderu, and San Communities of Namibia: Volume 3, University of Namibia Press, Windhoek, 570 p.

Hirshleifer, J., DeHaven, J.C., and Milliman, J.M., 1960, Water Supply – Economics, technology, and policy: University of Chicago Press, 1960, p 59–60. https://babel.hathitrust.org/cgi/pt?id=mdp.39015011501270&view=1up&seq=11.

Hiscock, K.M., M.O. Rivett, and R.M. Davison, 2002. Sustainable groundwater development: in Sustainable Groundwater Development: Special Publication No. 193. London: Geological Society, p 1–14.

Hjálmarsson, J.R., 2007, History of Iceland – From the Settlement to the Present Day: Forlagid, Reykjavik, 208 p.

Hopkins, D., 1987, Management of the Gnangara Mound groundwater resources [Western Australia]: International Commission on Irrigation and Drainage. Australian National Committee, Australian National Committee, p. 1.24–1.34.

Hovorka, S.D., Mace, R.E., and Collins, E.W., 1998, Permeability structure of the Edwards Aquifer, South Texas – Implications for aquifer management: Bureau of Economic Geology, The University of Texas at Austin, Report of Investigations No. 250, 55 p.

Hufschmidt, M., 1965, The methodology of water resource system design *in* Burton, I., and Kates, R., editors, *Readings in Resource Management and Conservation*: University of Chicago Press, p 558–570.

IUCN (International Union for Conservation of Nature and Natural Resources), 1980, *World Conservation Strategy – Living Resource Conservation for Sustainable Development*: International Union for Conservation of Nature and Natural Resources, United Nations Environment Programme, and the World Wildlife Fund; 77 p.

Janasie, C., 2020, An overview of water law in Illinois: Sea Grant Law Center, February 2020, NSGLC-20-04-02, 11 p.

Jansen, M., 1989, Water Supply and Sewage Disposal at Mohenjo-Daro: World Archaeology, v 21, n 2, p 177–192.

Jasechko, S., Perrone, D., Seybold, H., Fan, Y., and Kirchner, J.W., 2020, Groundwater level observations in 250,000 coastal US wells reveal scope of potential seawater intrusion: Nature Communications, 11:3229 | https://doi.org/10.1038/s41467-020-17038-2.

Joji, V.S., Viswanathan, S., and Sebastian, S., 2021, A textbook of managed aquifer recharge: Scholars' Press, KS Omniscriptum Publishing, 120 p.

Jokisch, A., Schulz, O., and Papangelou, A., 2018, Geographic conditions of central-northern Namibia *in* Liehr, S., Kramm, J., Jokisch, A., and Müller, K., editors, Integrated Water Resources Management in Water-Scarce Regions – Water Harvesting, Groundwater Desalination and Water Reuse in Namibia: IWA Publishing, p 10–18.

Jónsson, G.S., Jensson, H., Weisshappel, J.B., Þórðarsson, T., and Karlsdóttir, S.D., 2014, Water Framework Directive (2000/60/EC), Article 5 (and 6), Summary Report – Characteristics and impact analyses: Environment Agency of Iceland, 29 p.

Jung, L., and Fang, L.-C.T., 1948, An Account of the Salt Industry at Tzu-liu-ching Tzu-liu-ching chi: Isis, v 39, n 4, p 228–234.

Kalf, F.R.P., Woolley, D.R., 2005. Applicability and methodology of determining sustainable yield in groundwater systems. Hydrogeology Journal 13, 295–312.

Kazmann, R.G., 1951, The role of aquifers in water supply: Transactions of the American Geophysical Union, v 32, p 227–230.

Kazmann, R.G., 1956, "Safe yield" in ground-water development, reality or illusion? Journal of the Irrigation and Drainage Division, Proceedings of the American Society of Civil Engineers, Paper 1103, 12 p.

Kelley, V.A., Ewing, J., Jones, T.L., Young, S.C., Hamlin, S., Jigmond, M., Harding, J., Pinkard, J., Yan, T.T., Scanlon, B., Reedy, B., Beach, J., Davidson, T., and Laughlin, K., 2014, Updated groundwater availability model of the Northern Trinity and Woodbine aquifers: Final model report to the North Texas Groundwater Conservation District, Northern Trinity Groundwater Conservation District, Prairielands Groundwater Conservation District, and Upper Trinity Groundwater Conservation District; Volume 1, INTERA, variously paginated (990 total pages).

Ketilsson, J., Óskarsdóttir, S. M., Claesson, A., and Collett, N.J., 2017, The Kerauga Cave and Lækjarbotnaveita in South Iceland – Groundwater safety and hydrogeology: Journal of Water Security, v. 3, https://doi.org/10.15544/jws.2017.001.

Khan, L.R., and Mawdsley, J.A., 1988, Reliable yield of unconfined aquifers: Hydrological Sciences Journal, v33, no. 2, p. 151–171.

Kiernan, K., Wood, C., and Middleton, G., 2003. Aquifer structure and contamination risk in lava flows--Insights from Iceland and Australia: Environmental Geology. v. 43, no. 7, p/ 852-865 10.1007/s00254-002-0707-8

Kløve, B., Margrethe, H., Kvitsand, L., Pitkänen, T., Gunnarsdottir, T.J., Gaut, S., Gardarsson, S.M., Rossi, P.M., and Miettinen, I., 2017, Overview of groundwater sources and water-supply systems, and associated microbial pollution, in Finland, Norway and Iceland: Hydrogeology Journal, v. 25, p. 1033–1044.

Kluge, T., and Polak, M., 2018, Water supply infrastructure and sanitation in central-northern Namibia *in* Liehr, S., Kramm, J., Jokisch, A., and Müller, K., editors, Integrated Water Resources Management in Water-Scarce Regions – Water Harvesting, Groundwater Desalination and Water Reuse in Namibia: IWA Publishing, p 19–22.

Knobloch, S., Klonowski, A., and Marteinsson, V.Þ, 2020, Microbial diversity of the Heiðmörk groundwater system and potential sources of drinking water contamination: Skýrsla Matís 06–20, January, 40 p.

Knobloch, S., Klonowski, A., Tómasdóttir, S., Kristjánsson, B.R., Guðmundsson, S., and Marteinsson, V.Þ, 2021, Microbial intrusion and seasonal dynamics in the groundwater microbiome of a porous basaltic rock aquifer used as municipal water reservoir: FEMS Microbiology Ecology, v. 97, no. 3, https://doi.org/10.1093/femsec/fiab014.

Konikow, F. 2011, Contribution of global groundwater depletion since 1900 to sea-level rise: Geophysical Research Letters, v 38, 5 p.

Kresic, N., 2008, Groundwater Resources – Sustainability, Management, and Restoration: The McGraw-Hill Companies, Inc., 852 p.

Krutilla, J., and Eckstein, O., 1958, *Multiple Purpose River Development*: The Johns Hopkins Press, Baltimore, 336 p.

Lane, K., 2017, Puquios and aqueducts in the Central Andes of South America: Boca Raton, Florida: CRC Press. p. 465–467, in Underground Aqueducts Handbook.

Larsen, P., 2003, Scientific Accounts of a Vanishing Lake–Janez Valvasor, Lake Cerknica, and the New Philosophy: Master's Thesis, MSc in History of Science and Technology, Oxford University, 52 p.

Lawrence, C.R., 2018, A history of hydrogeology in Australia from pre-European to the 21st Century: Modern Environmental Science and Engineering, v 4, n 4, p 276–308.

le Demour, K., 1732, Machines approvées: L'Academie Royale des Sciences de Paris, v 6, n 363, p 9.

Leahy, T.C., 2016, Desperate times call for sensible measures – The making of the California Sustainable Groundwater Management Act: Golden Gate University School of Law Environmental Law Journal, v. 9, n. 1, p. 5–40.

Lee, C.H., 1912, An intensive study of the water resources of a part of Owens Valley, California: U.S. Geological Survey Water-Supply Paper 294, 144 p.

Lee, C.H., 1915, The determination of safe yield of underground reservoirs of the closed basin type: Transactions of the American Society of Civil Engineers, v 78, n 1, p 148–151.

Leib, D.I., and Stiles, T.C., 1998, Yield estimates for surface-water sources in Sophocleous, M., editor, Perspectives on Sustainable Development of Water Resources in Kansas, Kansas Geological Survey Bulletin 239.

Litton, A., 1860, Belcher & Brother's artesian well: Transactions of the Academy of Science of Saint Louis, v. 1, p. 80–86.

Loaiciga, H.A., 2016, The safe yield and climatic variability – Implications for groundwater management: Groundwater, v 55 n 3, p. 334–345.

Lohman, S.W., 1972, Ground-water hydraulics: U.S. Geological Survey Professional Paper 708, 70 p.

Loucks, D.P., 2000, Sustainable water resources management: Water International, v. 25, n. 1, p. 3–10, https://doi.org/10.1080/02508060008686793.

Lynch, P.N., Shepard, C.U., and Geddings, J.F.M., 1884, Artesian well – Charleston's water supply: The Sanitarian, Medico-Legal Society of New York, v. 12, p. 110–127.

Macdonald, A.M., Calow, R.C., Macdonald, D.M.J., Darling, W.G., and Dochartaigh, B.É.Ó., 2009, What impact will climate change have on rural groundwater supplies in Africa? Hydrological Sciences–Journal–des Sciences Hydrologiques, v 54, n 4, p 690–703.

Mace, R.E., 2016, So secret, occult, and concealed – An overview of groundwater management in Texas: in Law of the Rio Grande, Water Law Institute, CLE International, April 14–15, 2016, La Fonda, Santa Fe, New Mexico, Section H3, 27 p.

Mace, R.E., 2018, Groundwater Management: in Groundwater – State of the Science and Practice, National Ground Water Association Press, Westerville, Ohio, p. 38–40.

Mace, R.E., 2019, Groundwater governance, management, regulation, and emerging issues in the Lone Star State: Water Resources IMPACT, American Water Resources Association, p.11–12.

Mace, R.E., 2021a, Five gallons in a ten-gallon hat – Groundwater sustainability in Texas: The Meadows Center for Water and the Environment, Texas State University, Report 2021-08, 50 p.

Mace, R.E., 2021b, Stakeholder engagement in water resources decisions in Texas: American Association of Geographers Annual Meeting, April 7–11, 2021.

Mace, R.E., Dutton, A.R., and Nance, H.S., 1994, Water-level declines in the Woodbine, Paluxy, and Trinity aquifers of North-Central Texas: Gulf Coast Association of Geological Societies Transactions, v 44, p 413–420.

Mace, R.E., Leurig, S., Seely, H., and Wierman, D.A., 2020, Bringing back Comanche Springs – An analysis of the history, hydrogeology, policy, and economics: Report 2020-08, The Meadows Center for Water and the Environment, Texas State University, 153 p.

Mace, R.E., and Ridgeway, C.K., 2016, Forecasting underground rain – Groundwater availability modeling: in Sahs, M.K., editor, Essentials of Texas Water Resources, TexasBarBooks, State Bar of Texas, Chapter 19, p. 423–438.

Mace, R.E., Ridgeway, C., and Sharp, J.M., Jr., 2004, Groundwater is no longer secret and occult – A historical and hydrogeologic analysis of the East case: in Mullican, W. F., III, and Schwartz, S., editors, 100 Years of Rule of Capture: From East to Groundwater Management, Texas Water Development Board Report 361, p 63–88.

Mackworth, H., 1854, Improvements in boring for minerals, wells, and shafts: Transactions of the North of England Institute of Mining and Mechanical Engineers, v. 2.

Maclay, R.W., 1995, Geology and hydrology of the Edwards Aquifer in the San Antonio Area, Texas: U.S. Geological Survey, Water-Resources Investigations Report 95-4186, 64 p.

Maimone, M., 2004, Defining and managing sustainable yield: Ground Water, v 42, n 6, p 809–814.

Malick, T., Redford, R., Dunn, L., Sewell, D., Warren, W., Sewell, J., and Daniel, L., 2008, The Unforeseen: Two Birds Films, 92 minutes.

Maloszewski, P., Witczak, S., and Malina, G., editors, 2012, Groundwater Quality Sustainability: CRC Press, 370 p.

Mann, J.F., Jr., 1961, Factors affecting the safe yield of ground-water basins: Journal of the Irrigation and Drainage Division, Proceedings of the American Society of Civil Engineers, v 87, n 3, p 63–69.

Mann, J.F., Jr., 1963, Factors affecting the safe yield of ground-water basins: Transactions of the American Society of Civil Engineers, v 128, Paper No. 3434, p 180–186.

Mann, J.F., 1963, Response to comments on "Factors affecting the safe yield of ground-water basins:" Transactions of the American Society of Civil Engineers, v 128, Paper No. 3434, p 189–190.

Mann, J.T., Jr., 1968, Concepts in ground water management: Journal of the American Water Works Association, v 60, n 12, p 1336–1344.

Mapani, B.S., 2005, Groundwater and urbanisation, risks and mitigation – The case for the city of Windhoek, Namibia: Physics and Chemistry of the Earth, v 30, p 706–711.

Maréchal, J.-C., and Rouillard, J., 2020, Groundwater in France – Resources, use, and management Issues – Sustainable Groundwater Management A Comparative Analysis of French and Australian Policies and Implications to Other Countries, p.17–45, https://doi.org/10.1007/978-3-030-32766-8_2. hal-02512154.

Margat, J., and van der Gun, J., 2013, Groundwater Around the World – A Geographic Synopsis: CRC Press (Taylor & Francis Group), 341 p.

Marsh, B., and Marsh, A., 1993, He brought first bathtub to Fort Worth: The Tarrant County Historical Journal, Spring Edition.

MAWRD (Ministry of Agriculture, Water and Rural Development), 2000, National water policy white paper: Republic of Namibia, 45 p.

Mays, L.W., 2010, Water technology in ancient Egypt in Mays, L.W., editor, Ancient Water Technologies: Springer Science and Business Media. B.V.: Dordrecht, The Netherlands, Chapter 3, p 53–66.

McCarthy, J.J., Canziani, O.F., Leary, N.A., Dokken, D.J., and White, K.S., editors, 2001, Climate Change 2001 – Impacts, Adaptation, and Vulnerability: Cambridge University Press, 1032 p.

McKay, J., 2002, Encountering the South Australian landscape – Early European misconceptions and our present water problems: Hawke Institute Working

Papers Series, No. 21, Hawke Institute, University of South Australia, Magill, South Australia, 36 p.

McGuinness, C.L., 1951, The water situation in the United States with special reference to ground water: U.S. Geological Survey Circular No. 114, 127 p.

Megdal, S.B., Gerlak, A.K., Varady, R.G., and Huang, L.Y., 2015, Groundwater governance in the United States: Common priorities and challenges: Groundwater, v 53, n 5, p 677–684. https://doi.org/10.1111/gwat.12294.

Meinzer, O.E., 1915, Discussion of "The determination of safe yield of underground reservoirs of the closed basin type:" Transactions of the American Society of Civil Engineers, v 78, n 1, p 234–239.

Meinzer, O.E., 1920, Quantitative methods of estimating ground-water supplies: Bulletin of the Geological Society of America, v 31, p 329–338.

Meinzer, O.E., 1923, Outline of ground-water hydrology with definitions: U.S. Geological Survey Paper 494, 71 p.

Meinzer, O.E., 1932, Outline of Methods for Estimating Ground-Water Supplies: U.S. Geological Survey Water-Supply Paper No. 638-C, p 99–144.

Meinzer, O.E., 1934, History and development of ground-water hydrology: Journal for the Washington Academy of Sciences, v 24, n 6, p 6–32.

Merriam-Webster, 2021a, History and Etymology for well: https://www.merriam-webster.com/dictionary/well.

Merriam-Webster, 2021b, Maintain: in Merriam-Webster.com dictionary. Retrieved September 30, 2021, from https://www.merriam-webster.com/dictionary/maintain.

Merriam-Webster, 2021c, Permeable: Merriam-Webster.com Dictionary, https://www.merriam-webster.com/dictionary/permeable, accessed March 23, 2021.

Meunier, M., 1978, Les puits artésiens de l'Avallonnais: Académie du Morvan, v. 5, n. 6, p. 3–14. https://gallica.bnf.fr/ark:/12148/bpt6k96033964/f5.item.r=grenelle%20puit.

Miller, G.R., 1973, Shaping California water law – 1781 to 1928: Southern California Quarterly, v. 55, n. 1, p. 9–42.

Milliman, J.W., 1956, Commonality, the price system, and use of water supplies: Southern Economic Journal, v 22, n 4, p 426–437.

Milman, A., and Kiparsky, M., 2020, Concurrent governance processes of California's Sustainable Groundwater Management Act: Society & Natural Resources, v. 33, n. 12, p. 1555–1566.

Molle, F., 2011, Aquifer safe yield – Hard science or boundary concept? Ground Water 2011 Conference, Orléans (FRA), 2011/03, 2 p.

Morris, B.L., Lawrence, A.R.L., Chilton, P.J.C., Adams, B., Calow, R.C., and Klinck, B.A., 2003, Groundwater and Its Susceptibility to Degradation – A

Global Assessment of the Problem and Options for Management: United Nations Environment Programme, Early Warning and Assessment Report Series, 03-3, 126 p.

Muckel, D.C., 1955, Pumping ground water so as to avoid overdraft: Yearbook of Agriculture 1955, U.S. Department of Agriculture, p 294–301.

Mudd, G.M., 2000, Mound springs of the Great Artesian Basin in South Australia – A case study from Olympic Dam: Environmental Geology, v 39, n 5, p 463–476.

NGC (National Groundwater Committee), 2004, Knowledge gaps for groundwater reform: in Proceedings from Workshop 12–13, November 2003, Canberra, Australia.

Nhuleipo, O., 2022, Namibia: Gaborone Declaration for Sustainability in Africa, accessed April 24, 2022. http://www.gaboronedeclaration.com/namibia.

Nicol, S., 2005, The Great Artesian Basin: Department of Natural Resources and Water, The State of Queensland, accessed April 10, 2022, https://web.archive.org/web/20070315203118/http://www.nrw.qld.gov.au/water/gab/.

Norton, W.H., 1897, Artesian wells of Iowa: Iowa Geological Survey, v 6, 130 p.

NRMMC (Natural Resource Management Ministerial Council), 2002, Managing overallocated groundwater systems: Commonwealth of Australia.

O'Riordan, T., 2000, The sustainability debate *in* O'Riordan, T., editor, *Environmental Science for Environmental Management*, Prentice Hall, Harlow, Essex, p 29–62.

Olafsson, S., Cook, D., Davidsdottir, B., and Johannsdottir, L., 2014, Measuring countries' environmental sustainability performance – A review and case study of Iceland: Renewable and Sustainable Energy Reviews 39934-948 S136403211400553X 10.1016/j.rser.2014.07.101

Orkustofnun, 2021, Orkustofnun Data Repository OS-2021-T009-01.

OUP (Oxford University Press), 2021, Oxford English Dictionary: https://www.oed.com/.

Owen, J., 1915, Discussion of "The determination of safe yield of underground reservoirs of the closed basin type:" Transactions of the American Society of Civil Engineers, v 78, n 1, p 219–221.

Oxford University Press, 2021, Policy: Oxford English Dictionary, accessed November 30, 2021, https://www.oed.com/view/Entry/146842?result=1&rskey=YjloBU&.

Palmer, R.L., 1965, Ground water – Depletion of a wasting asset: Southwestern Law Journal, December, v 19, p 791–800.

Papin, D., 1689, Rotatilis suctor et Pressor Hassiacus, in serenisfima aula caffellana demonstratus & detectus: Actis Eruditorum, June, p 317–322.

Parker, G.G., 1951, Geologic and hydrologic factors in the perennial yield of the Biscayne Aquifer: Journal of the American Water Works Association, v 43, p 817–834.

Parkinson, C.E., 1954, Minority Report *in* Artesian Water Investigation Committee, Artesian water supplies in Queensland: Queensland Artesian Water Investigation Committee, Department of the Co-ordinator-General of Public Works, Queensland, p 70–79.

Paterson, A.B., 1902, Rio Grande's last race and other verses: Angus and Robertson, Melbourne.

Pavley, F., Dickinson, R., and Rendon, A., 2014, Sustainable groundwater management: Senate Bill No. 1168, http://www.leginfo.ca.gov/pub/13-14/bill/sen/sb_1151-1200/sb_1168_bill_20140822_amended_asm_v92.htm

Peltenburg, E.J., Croft, P., Jackson, A., McCartney, C., and Murray, M.A., 2001, Well-Established Colonists: Mylouthkia 1 and the Cypro-Pre-Pottery Neolithic B in Swiny, S., (editor), The Earliest Prehistory of Cyprus: From Colonization to Exploitation, Swiny, American Schools of Oriental Research Archaeological Reports 5; Cyprus American Archaeological Research Institute Monograph Series 2. Boston: American Schools of Oriental Research, p 61–93.

Peralta, R., Timani, B., and Das, R., 2011, Optimizing safe yield policy implementation: Water Resources Management, v 25, p 483–508.

Pierce, S.A., Sharp, Jr., J.M., Guillaume, J.H.A., Mace, R.E., and Eaton, D.J., 2012, Aquifer-yield continuum as a guide and typology for science-based groundwater management: Hydrogeology Journal, v 21, p 331–340.

Pipkin, T., 1993, Barton Springs – The soul of a city: Softshoe Pub, 135 p.

Pittman, E.F., and David, T.W.E., 1903, Irrigation geologically considered with special reference to the artesian area of New South Wales: Journal and Proceedings of the Royal Society of NSW, v 37, p CIII–CLIII.

Polak, M., 2014, Zwischen Haushalt und Staat – Lokale Water Governance im zentralen Norden Namibias (Between Household and State – Local Water Governance in Central Northern Namibia): Forum Politische Geographie, Book 12, LIT-Verlag Münster, Berlin, 240 p.

Polak, M., and Kluge, T., 2018, Water governance in Namibia – Progress in and challenges of putting IWRM into practice *in* Liehr, S., Kramm, J., Jokisch, A., and Müller, K., editors, Integrated Water Resources Management in Water-Scarce Regions – Water Harvesting, Groundwater Desalination and Water Reuse in Namibia: IWA Publishing, p 23–26.

Powell, O., Silcock, J., and Fensham, R., 2015, Oases to oblivion – The rapid demise of springs in the South-Eastern Great Artesian Basin, Australia: Groundwater, v 53, n 1, p. 171–178.

Price, M., 2002, Who needs sustainability? *in* Hiscock, K.M., Rivett, M.O., and Davison, R.M., editors, *Sustainable Groundwater Development*: Geological Society, London, Special Publications, 193, p 75–82.

Pyarelal, N., 1956, Mahatma Gandhi – The Last Phase: Navajivan Publishing House, India, 750 p.

Pyne, D.G., 2019, Groundwater recharge and wells – A guide to aquifer storage recovery: CRC Press, 400 p.

Ravenstein, E.G., 1900, The voyages of Diogo Cão and Bartholomeu Dias, 1482–88: The Geographic Journal, variously paginated.

Redacted and Redacted, 2016, Discount rates in the economic evaluation of U.S. Army Corps of Engineers Projects: CRS Report prepared for members and committees of Congress, Congressional Research Service, R44594, 37 p.

Remmert, D., 2016, Water governance in Namibia – A tale of delayed implementation, policy shortfalls, and miscommunication: Special Briefing Report No. 13, Democracy Report, 20 p.

Renshaw, E.F., 1963, The management of ground water reservoirs, Journal of Farm Economics, v 45, 285–295.

Rinaudo, J.-P., 2020, Groundwater policy in France: From private to collective management in Rinaudo, J.-P., Holley, C., Barnett, S., and Montginoul, M., editors, Sustainable Groundwater Management – A Comparative Analysis of French and Australian Policies and Implications to Other Countries: Chapter 3, Global Issues in Water Policy, Volume 24, Springer, p. 47–65.

Robinson, Z.P., Fairchild, I.J., and Russell, A.J., 2008, Hydrogeological implications of glacial landscape evolution at Skeiðarársandur, SE Iceland: Geomorphology, v. 97, p. 218–236.

Rohrbough, M.J., 2013, Rush to Gold – The French and the California Gold Rush, 1848–1854: Yale University Press, 352 p.

Rudestam, K., and Langridge, R., 2014, Sustainable yield in theory and practice – Bridging scientific and mainstream vernacular: Groundwater, v. 52, p. 90–99.

Rybníčeka, M., Chlupc, T., Kalábekd, M., Kalábkováe, P., Kočárf, P., Kynclg, T., Muiggh, B., Tegelh, W., Vostrovskáe, I., and Kolářa, T., 2018, New dendroarchaeological evidence of water well constructions reveals advanced Early Neolithic craftsman skills: Dendrochronologia, v 50, p 98–104. https://doi.org/10.1016/j.dendro.2018.06.003.

Salk, J., 1992, Are We Being Good Ancestors? World Affairs: The Journal of International Issues, v. 1, n. 2, p. 16–18.

Sato, S., 1966, Ground water rights and depletion deduction: Natural Resources Journal, v 6, n 2, p 237–247.

Sax, J.L., 2002, We don't do groundwater – A morsel of California legal history: University of Denver Water Law Review 6:269.

Scanlon, B.R., Mace, R.E., Barrett, M.E., and Smith, B., 2003, Can we simulate regional groundwater flow in a karst system using equivalent porous media models? Case study, Barton Springs Edwards aquifer, USA: Journal of Hydrology, v. 276, n. 1, p. 137–158. https://doi.org/10.1016/S0022-1694(03)00064-7.

Schnegg, M., Bollig, M., and Linke, T., 2016, Moral equality and success of common-pool water governance in Namibia: Ambio, v 45, p 581–590. https://doi.org/10.1007/s13280-016-0766-9.

Schneider, G.I.C., Schneider, M.B., and du Pisani, A.L, 2001, Natural and Socio-Economic Environment *in* Christelis, G., and Struckmeier, W., Groundwater in Namibia – An explanation to the hydrogeological map: Ministry of Agriculture, Water, and Rural Development, Namibia, p 14–29.

Scott, A., 1955, *Natural Resources – The Economics of Conservation*: University of Toronto Press, Toronto, 184 p.

Sellers, J.H., 1973, Tax implications of ground-water depletion: Ground Water, v 11, n 6, p 27–35.

Schenk, H.J., and Jackson, R.B., 2002, Rooting depths, lateral root spreads and below-ground/above-ground allometries of plants in water-limited ecosystems: Journal of Ecology, v 90, p. 480–494.

Shepherd, D.G., 1990, Historical development of the windmill: NASA Contractor Report 4337, 43 p.

Shiklomanov, I., 1993, World fresh water resources in Gleick, P.H., editor, Water in Crisis – A Guide to the World's Fresh Water Resources: Oxford University Press, p 13–24.

Shufeldt, G.A., 1867, History of the Chicago Artesian Well – A demonstration of the truth of the spiritual philosophy with an essay on the origin and uses of petroleum: Religio-Philosophical Publication Association, 58 p.

Siebert, S., Burke, J., Faures, J.M., Frenken, K., Hoogeveen, J., D€oll, P., and Portmann, F.T., 2010, Groundwater use for irrigation – A global inventory: Hydrology and Earth System Sciences, v 14, p 1863–1880.

Sigurdsson, F., 1987, Hydrogeology and geohydrology: Report 6, National Energy Authority, Reykjavik, Iceland, 28 p. + 21 plates.

Simmons, C.T., 2003, Happy 200th birthday Mr Darcy and our thanks for your law! A tribute editorial celebrating the life and times of the father of our science, Henry Darcy (1803–1858): Hydrogeology Journal, v 11, p. 611–614.

Smith, G.E.P., 1915, Discussion of "The determination of safe yield of underground reservoirs of the closed basin type:" Transactions of the American Society of Civil Engineers, v 78, n 1, p 221–234.

Smith, G.E.P., 1924, The effect of transpiration of trees on the ground-water supply (unpublished paper given before the Geological Society of Washington, Nov. 22, 1922): see Journal of the Washington Academy of Sciences, v. 14, no. p. 160.

Smith, M., Cross, K., Paden, M. and Laban, P., editors, 2016, Spring – Managing groundwater sustainably: International Union for Conservation of Nature, Gland, Switzerland, 133 p.

Snyder, J.H., 1955, Ground water in California – The experience in Antelope Valley: Giannini Foundation Ground Water Studies No. 2, Agricultural Experiment Station, University of California, 171 p.

Soetaert, F., Wanke, H., Dupuy, A., Lusuekikio, V., Gaucher, E.C., Bordmann, V., Fleury, J.-M., and Franceschi, M., 2022, Toward the sustainable use of groundwater springs – A case study from Namibia: Sustainability, v 14, 3995. https://doi.org/10.3390/su14073995.

Solanes, M., and Gonzalez-Villarreal, F., 1999, The Dublin Principles for water as reflected in a comparative assessment of institutional and legal arrangements for Integrated Water Resources Management: TAC Background Papers No. 3, Technical Advisory Committee, Global Water Partnership, 43 p.

Sophocleous, M., 1997, Concluding comments on managing water-resources systems – Why "safe yield" is not sustainable in Perspectives on Sustainable Development of Water Resources in Kansas, Kansas Geological Survey Bulletin 239, Chapter 10, p. 211–213.

Sophocleous, M., 2000, From safe yield to sustainable development of water resources – The Kansas experience. Journal of Hydrology, v 235, n 1–2, p 27–43.

Sopp, C.W., 1946, Discussion on "Utilization of groundwater storage in stream system development:" Transactions of the American Society of Civil Engineers, v 111, n 1, p 326–330.

SOSA (Save Our Springs Alliance), 2022, The story of the Save Our Springs Alliance: accessed January 4, 2022; https://www.sosalliance.org/our-story.html.

Stanford, J.D., Hemingway, R., Rohling, E.J., Challenor, P.G., Medina-Elizalde, M., and Lester, A.J., 2011, Sea-level probability for the last deglaciation – A statistical analysis of far-field records: Global and Planetary Change, v. 79, n. 3, p. 193–203. https://doi-org.libproxy.txstate.edu/10.1016/j.gloplacha.2010.11.002.

Steinel, A.T., 1926, History of Agriculture in Colorado – A Chronological Record of Progress in the Development of General Farming, Livestock

Production and Agricultural Education and Investigation, on the Western Border of the Great Plains and in the Mountains of Colorado, 1858 to 1926: Colorado Agricultural College, Fort Collins, 659 p.

Stephens, D.B., 1996, Vadose Zone Hydrology: CRC Lewis Publishers, 347 p.

Streit, K., and Garfinkel, Y., 2015, A specialized ceramic assemblage for water pulling – The Middle Chalcolithic well of Tel Tsaf, Israel: Bulletin of the American Schools of Oriental Research, v 374, p 61–73.

Stuart, W.T., 1945, Groundwater resources of the Lansing area: Michigan Geological Survey Progress Report 13, 35 p.

Suter, M., 1963, Discussion of "Factors affecting the safe yield of ground-water basins": Transactions of the American Society of Civil Engineers, v 128, Paper No. 3434, p 188–189.

Swindell, J.G., and Burnell, G.R., 1883, Rudimentary treatise on wells and well-sinking: Crosby Lockwood and Co., London, 166 p.

SWRCB (State Water Resources Control Board), 2013, Groundwater Workplan Concept Paper – Discussion Draft: October 4th, http://www.waterboards.ca.gov/water_issues/programs/groundwater/docs/gw_workplan100713.pdf.

Taghavi-Jeloudar, M., Han, M., Davoudi, M., Kim, M., 2013, Review of ancient wisdom of qanat, and suggestions for future water management: Journal of Environmental Engineering and Science, v 18, n 2, p 57–63.

Theis, C.V., 1935, The relation between the lowering of the piezometric surface and the rate and duration of discharge of a well using groundwater storage: Transactions of the American Geophysical Union, v 16, p 519–524.

Theis, C.V., 1940, The source of water derived from wells – Essential factors controlling the response of an aquifer to development: Civil Engineer, v 10, p 277–280.

Thomas, H.E., 1951, *The Conservation of Groundwater*: McGraw-Hill Book Company, New York.

Thomas, H.E., 1955, Water rights in areas of ground-water mining: U.S. Geological Survey Circular 347, 16 p.

Thomas, R.O., 1959, Legal aspects of ground water utilization: Proceedings of the American Society of Civil Engineers, v 85, n 4, p 41–63.

Thomas, R.O., 1963, Discussion of "Factors affecting the safe yield of ground-water basins": Transactions of the American Society of Civil Engineers, v 128, Paper No. 3434, p 186–188.

Thompson, Jr., B.H, M. Rohde, Howard, J.K., and Matsumoto, S., 2021, Mind the gaps – The case for truly comprehensive sustainable groundwater management: Water in the West. Stanford Digital Repository. Available at: https://purl.stanford.edu/hs475mt1364.

Thornsteinsson, T., and Elíasson, J., 1970, Geohydrology of the Laugarnes Hydrothermal System in Reykjavík, Iceland: National Energy Authority of Iceland, March 1970, 29 p. + 11 figures.

Todd, D.K., 1959, *Groundwater Hydrology*: John Wiley and Sons, Inc., New York, 336 p.

Todd, D.K., and Meyer, C.F, 1971, Hydrology and geology of the Honolulu Aquifer: Proceedings of the American Society of Civil Engineers, Journal of the Hydraulics Division, v. 97. No. 2, p. 233–256.

Toll, N., Fratesi, S.B., Green, R.T., Bertetti, F.P., and Nunu, R., 2017, Water resource management of the Devils River watershed, final report: Southwest Research Institute report under contract to the Devils River Conservancy, at http://www.devilsriverconservancy.org/research/, accessed January 2022.

Tolman, C.F., 1937, *Ground Water*: McGraw-Hill Book Co., Inc., New York, N.Y., 1937, 593 p.

Tóth, J., 1963, A theoretical analysis of groundwater flow in small drainage basins: Journal of Geophysical Research, v. 68, n. 16. P. 4795–4812.

Tóth, J., 1973, Hydrogeology and yield evaluation of a municipal well field, Alberta, Canada: Hydrological Sciences Journal, v. 18, no. 2, p. 165–189.

Turral, H., and Fullagar, I., 2007, Institutional Directions in Groundwater Management in Australia *in* Giordano, M., and Villholth, K.G., editors, The agricultural groundwater revolution – Opportunities and threats to development: CAB International, p. 320–361.

TWDB (Texas Water Development Board), 1968, The Texas water plan: Texas Water Development Board, Austin, Texas; variously paginated (252 total pages).

Umenne, S., and Chirwa, V., 2012, Culture and environment as generators of sustainable design – The case of Hoachanas Village, Hardap Region, Namibia: Sustainable Futures: Architecture and Urbanism in the Global South Kampala, Uganda, June 27–30, p 167–176.

United Nations, 1992, Agenda 21: United Nations Conference on Environment & Development, Rio de Janeiro, Brazil, 3 to 14 June 1992, 351 p.

USFWS (United States Fish & Wildlife Service), 2021, Endangered Species Act – A history of the Endangered Species Act of 1973: accessed December 28, 2021: https://www.fws.gov/endangered/laws-policies/esa-history.html.

Utton, A.E., 1982, The development of international groundwater law: Natural Resources Journal, v 22, n 1, p 95–118.

Van Esch, R.P.M., 1997, Simulation of three-dimensional unsteady flow in hydraulic pumps: Ph.D. dissertation, University of Twente, Enschede, Nederlands, 147 p.

van Rensburg, P., 2018, Mitigation through innovation – Sustainable water supply in Windhoek: Urbanet, posted March 22, 2018; accessed April 24, 2022. https://www.urbanet.info/mitigation-through-innovation/.

Vernoux,·J.F., 2004, Réseau piézométrique du bassin Seine-Normandîe, Stations de surveillance de ta nappe de l'Albîen: BRGM, Report BRGM/RP-53009-FR, variously paginated (33 total pages).

Villholth, K.G., and Conti, K.I., 2019a, Groundwater governance – Rationale, definition, current state and heuristic framework: *in* Villholth, K.G., López-Gunn, E., Conti, K.I., Garrido, A., and van der Gun, J., Advances in Groundwater Governance: Taylor and Francis Group, p 3–31.

Villiers, Marq de, 2003, Water, Random House, 464 p.

Vitruvius (Marcus Vitruvius Pollio), n.d. (published ~30-15 BC), De Architectura, Book X.

Votteler, T.V., 1998, The little fish that roared – The Endangered Species Act, state groundwater law, and private property rights collide over the Texas Edwards Aquifer: Environmental Law, v. 28, n. 4, p. 845–879.

Voudouris, K., Valipour, M., Kaiafa, A., Zheng, X.Y., Kumar, R., Zanier, K., Kolokytha, E., and Angelakis, A., 2019, Evolution of water wells focusing on Balkan and Asian civilizations: Water Supply, v 19, n 2, p 347–364.

Wallace, M., 2011, A history of Namibia – From the beginning to 1990: Columbia University Press, New York, 451 p.

Walton, W.C., 1964, Future water-level declines in deep sandstone wells in Chicago region: Ground Water, v 2, n 1, p 13–20.

Watson, A., Mommsen, T., and Krueger, P., 1985, The Digest of Justinian: University of Pennsylvania Press; Philadelphia, Pennsylvania.

WCED (World Commission on Environment and Development), 1987, *Our Common Future*. New York: Oxford University Press, 300 p.

Weinberg, A., French, L.N., and Perez, J.B., 2018, Overview of groundwater conditions in Val Verde County, Texas: Texas Water Development Board, 227 p.

Weiner, J., 1994, Well on my back – An update on the Bandkeramik wooden well of Erkelenz-Kückhoven: NewsWARP, v 16, p 5–17.

Wells, D.A., Cross, C.R., Bliss, G., Trowbridge, J., Kneeland, S., and Nichols, R., 1854, Annual of scientific discovery – Or, year-book of facts in science and art: Gould and Lincoln.

Wenzell, H.B., 1909, Ed. Erockson v. Crookston Waterworks, Power, & Light Company: Minnesota Reports – Cases Argued and Determined in the Supreme Court of Minnesota, v. 105, p. 182–198.

Whitney, J.D., 1894, The United States – Facts and figures illustrating the physical geography of the country, and its material resources: Supplement I. Population: Immigration: Irrigation, Little, Brown, and Company, Boston, 324 p.

Wilkinson, C.S., 1882, Notes on the occurrence of artesian wells in the Albert District, New South Wales: Proceedings of the Linnean Society of NSW, n 6, p 155–157.

Wilkinson, T.J., Boucharlat, R., Ertsen, M.W., Gillmore, G., Kennet, D., Magee, P., Rezakhani, K., and Schacht, T., 2012, From human niche construction to imperial power – Long-term trends in ancient Iranian water systems: Water History, v 4, p 155–176.

Williams, C.C., and Lohman, S.W., 1949, Geology and ground-water resources of a part of south central Kansas: Geological Survey of Kansas, Bulletin 79, 212 p. http://www.kgs.ku.edu/Publications/Bulletins/79/index.html.

Williamson, D.H., 1996, The development of hydrogeology in Australia: self-published book, Ermington, N.S.W., ISBN 0646411683, 50 p.

Williamson, W.H., 2012, The history of hydrogeology in Australia *in* Howden, N., and Mather, J., editors, History of Hydrogeology, CRC Press, p 7–27.

Wise, S.M., 1996, The legal thinghood of nonhuman animals: Boston College of Environmental Affairs Law Review, v 23, n 3, p. 471–547.

Wolff, H.C., 1911, The utilization of the underflow near St. Francis, Kansas: U.S. Geological Survey Water-Supply Paper 258, p. 104–107.

Wood, W.W., 2001, Water sustainability – Science or management? Ground Water, v 39, n 5, p 641.

Woolley, D.R., Habermehl, M.A., Palmer, J., Sibenaler, X., and O'Neill, D., 1987, Improved Efficiency of Artesian Groundwater Use, Rehabilitation of Bores and Conversion from Drains to Piping: Report to the Australian Water Resources Council Groundwater Committee by the Great Artesian Basin Sub-Committee, 18 p.

Worster, D., 1993, The shaky ground of sustainability *in* Sachs, W., editor, Global Ecology – A New Era of Political Conflict: Zed Books, p 133–145.

Yannopoulos, S.I., Lyberatos, G., Theodossiou, N., Li, W., Valipour, M., Tamburrino, A., and Angelakis, A.N., 2015, Evolution of water lifting devices (pumps) over the centuries worldwide: Water, v 7, p 5031–5060; https://doi.org/10.3390/w7095031.

Young, R.A., 1970, Safe yield of aquifers – An economic reformulation: Proceedings of the American Society of Civil Engineers, Journal of the Irrigation and drainage Division, v. 96, n. 4, p. 377–385.

Zhou, Y., 2009, A critical review of groundwater budget myth, safe yield, and sustainability: Journal of Hydrology, v 370, p. 207–213.
Zuber, A., Maloszewski, P., Witczak, S., and Malina, G., 2021, Groundwater quality sustainability: CRC Press, 364 p.

References

Cambridge Dictionary, 2021b, Policy: https://dictionary.cambridge.org/dictionary/english/policy accessed October 1, 2021.
Villholth, K.G., and Conti, K.I., 2019b, Groundwater governance—Rationale, definition, current state and heuristic framework: *in* Villholth, K.G., López-Gunn, E., Conti, K.I., Garrido, A., and van der Gun, J., Advances in Groundwater Governance: Taylor and Francis Group, p 3–31.

Index[1]

[1] Note: Page numbers followed by 'n' refer to notes.

R. E. Mace, *Groundwater Sustainability*, Palgrave Studies in Environmental Sustainability, https://doi.org/10.1007/978-3-031-13516-3

GPSR Compliance
The European Union's (EU) General Product Safety Regulation (GPSR) is a set of rules that requires consumer products to be safe and our obligations to ensure this.

If you have any concerns about our products, you can contact us on

ProductSafety@springernature.com

In case Publisher is established outside the EU, the EU authorized representative is:

Springer Nature Customer Service Center GmbH
Europaplatz 3
69115 Heidelberg, Germany

www.ingramcontent.com/pod-product-compliance
Lightning Source LLC
Chambersburg PA
CBHW062145170326
41400CB00005B/2736
9783031135156